JOHN HAY

The Immortal Wilderness

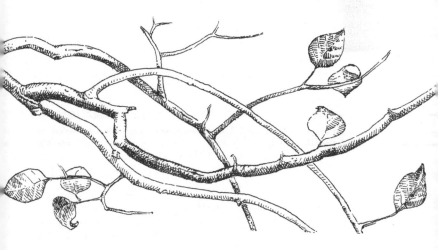

W · W · Norton & Company · New York · London

Excerpt on page 77
reprinted from
Richard K. Nelson,
Make Prayers to the Raven
copyright © 1983, with the kind permission of
The University of Chicago Press.

The text of this book is composed in Janson, with
display type set in Willow. Composition and
manufacturing by the Maple-Vail Book Manufacturing Group.
Title page illustration by Ann Bliss.

First Edition

Library of Congress Cataloging-in-Publication Data

Hay, John, 1915–
The immortal wilderness.
1. Natural history.   2. Nature conservation.
I. Title.
QH45.5.H375   1987   508   86–7119
ISBN 0-393-02385-0

W. W. Norton & Company, Inc.
500 Fifth Avenue, New York, N. Y. 10110
W. W. Norton & Company Ltd.
37 Great Russell Street, London WC1B 3NU

2 3 4 5 6 7 8 9 0

BOOKS BY JOHN HAY

A Private History (Poems)
The Run
Nature's Year
A Sense of Nature (with Arlene Strong)
The Great Beach
The Atlantic Shore (with Peter Farb)
In Defense of Nature
The Primal Alliance: Earth and Ocean
Spirit of Survival
The Undiscovered Country
The Immortal Wilderness

The Immortal Wilderness

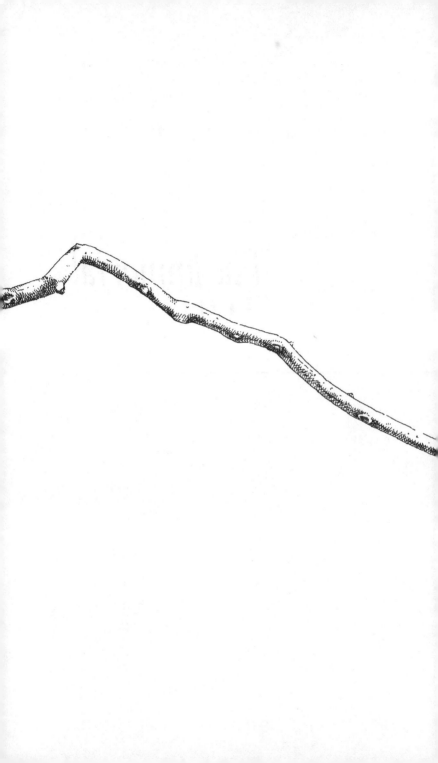

*To my Mother
For her love
And belief in courage*

Acknowledgments

Grateful acknowledgment is made to the following publications for sections or paragraphs of this book that appeared in their pages: *Orion; Defenders of Wildlife; The New England Review; Deep Ecology*, edited by Michael Tobias: *The Cape Naturalist; Diversion; Campus*. I would also like to thank Jake Page and his wife Suzanne for their company and encouragement.

Contents

Preface

Perhaps we all start with "nature" as a way to explain the surrounding world to ourselves. It is something to count on, to love or hate, and to exploit to our advantage, but in our time the term has been so monopolized, so twisted and worried to death, that it has almost lost its value. What is it we are talking about? Can "mother nature" be our protector any longer when we cast her out? We continually deny her in favor of our own ability to conquer, and even supersede her. This leads to inadmissible isolation. We do not know what to do with our runaway power.

The illusion grows, catastrophic in its implications, since it begins to remove us from land, water, and basic nourishment, that all around us lies a nature to which we are no longer obligated. Disengagement tempts us to brutalize the earth that sustains us. The lasting interdependence of people and land is denied, or at least obscured, and what is left of our roots surfaces through violent impulses only dimly aware of what motivates them. At the same time, the foundation of all life, which we never made, moves its mountains and continents over vast stretches of time and shakes the mercurial amalgam of water and light before our eyes. It insists on a universal engagement in which there is no distinction under the sun between human and "nonhuman" forms of life, and permits no self-made exiles.

New inventions, new worlds, come and go in an age of industrial and technological advance. Permanence and stability seem to be abandoned. I, for one, never meant to travel so far and fast as to leave all home and continuity behind me, but history, that great expropriator, travels with me. Human history is not new, of course, but very old, full of triumph, passion, and deceit. Mankind goes it alone, taking the chaotic risk of its pretensions, counting on a future it is unable to see. But at any period, we are still governed by a planetary scale of unfathomed darkness and a glory of changing light, by the limits put on us by ice and fire, drought and flood, and by those unrelenting tides where the marine fishes circle on. No triumph of mathematics and abstract thought can elevate us above the whole company of existence. It is the common performance, the magnitude of life's bending and rising like seabirds to the pressure of the elements, that constitutes permanence, and the earth's supreme integrity.

I have used the term *wilderness* in my title, and in the following pages, not to imply what we have "conquered" and passed by, or once associated with deserted and uninhabitable regions where religious spirits fought with their demons, but to mean the great container of life and death, the earth's immortal genius. Nothing in wilderness escapes the universal interdependence. Nothing in it lacks response to the rhythms of the planet in the skies. Through their sentient and perceptive properties, the greater company of life shares in that capacity. Language does not make us superior to all others; it originated in the earth's expressive needs. Nor do we have a monopoly on intelligence. It is not to deny the complex nature of human self-consciousness and the reaches of the spirit to suggest that those infinite sources humanity has called upon to prove it also nourish and sustain the rest of life. Millions of species share in the great unconscious sea, the shelter of vision and desire. Despite an age that seems to be casting itself loose from its moorings, those multiple voices and alliances are still addressing us, waiting to be heard. The true space in us is not finite; it does not depend on all the goods and conveniences, the momentum and machinery, the amplified distractions of our world to define it. It is open ended.

I am not satisfied that we can claim universal space as our own by peering hundreds of millions of miles into it through a telescope. I do not believe that we will ever discover the last thing about the properties of living matter by defeating the invisible with a microscope. I am obligated to science for its methods. They help us amateurs know what it is we are looking at; but there is a point at which science leaves off, and you have to trust the lasting wilderness receptor in yourself.

On the American continent, wilderness was quickly turned into a wrecker's paradise, often confused with that ideal "garden" which was to replace all that was wild and uncontrolled. Many who retain a sense of that original magnificence look back on it with some nostalgia. Yet only as an ideal, now vanishing in the distance, is it too easy to ignore. Its primal laws, as anyone contemplating the branching arrangements of a tree, or the flight of a bird, will understand, still reign, and we cannot assume we are not dependent on them. The idea of an absentee wilderness is a denial of our origins and, in fact, of the ground beneath our feet.

Exposure to the size and scale of the American continent made millions of immigrants feel not only embattled, but isolated, exposed, and even accidental. They did everything they could to build against that sense of estrangement, until the surrounding world felt safe, and in their hands. The almost total occupation, and therefore the desertion, of space, in other words, the mutual balanced existence of land and sea, left us with no guidance but our own. Wilderness was never accepted as our home, but now it has to be. Uncompromising but protective, it holds to the principles of renewal and diversity in all the facets of its nature. It keeps the law, for which we have no substitute. As original creation, it reconciles extremes in a way that is impossible for mankind to imitate. We know this, not as the outsiders and pretenders to the throne of nature that we appear to be, but in our unfinished selves, where stability and instability keep company with the eternal weather.

The Immortal
Wilderness

CHAPTER 1

Main Street on the Moon

Following World War ll, when it seemed that we had all been flung out, dispersed across the face of the earth, I was glad to be able to return to where I started from, but I was no longer quite sure where that was. With a half-expressed fear that I might become a displaced person in my own country, I set out to try and find a recoverable past. Wherever I found it, it seemed threatened, like those threatened and endangered species we have become aware of. One of my trips, in part accidental, because of a friend who had a temporary assignment in the press office there, was to Los Alamos, fountainhead of nuclear invention. It was 1950, and I set off from Texas in a bus full of sailors, air force personnel, and GIs on their way to California, still spreading out, although the war was over, to stations in the Pacific. I remember seeing my first wild turkey from the window of the bus. It had been standing by the side of the road and only just escaped as we roared by. It was a skinny, ungainly bird, not at all the great bronze, iridescent creature painted by Audubon; and it was ignominiously swept aside, to land awkwardly in the scrub, like a newspaper flung by the wind against a bench in the city park . . . fairly typical, I guess, of the way we had always treated original Americans.

The bus rolled across a far reaching land covered with cactus and feathery green mesquite. There were a few Mexicans on board, one of them singing a low song in Spanish, and three or four silent Texans, who disembarked at nameless places along the route to head for unseen houses in an empty land. After a while, the stars came out, traveling their immense distances overhead. Now and then an oil refinery materialized out of the growing darkness, fantastic cities with piercing lights, great silvery spheres, and tall slim tanks standing together. Off on the perimeter, pipes displayed their orange hands of burning gas.

The white, concrete highway stretched ahead from one rolling plain to the next. In spite of the barbed wire fences that had long since enclosed the open range, space was intoxicating. One could go riding out forever. That we should become an indoor people seemed like a betrayal of origins.

Later on, I changed buses and headed for New Mexico, "The Land of Enchantment," as it said on the license plates, an ancient land of dry seas crossed by lavender and gray shadows, of strange, twisted rocks, and little dark cedars holding on to gravelly slopes. And there were real enchantments, like the rainbow, Indian symbol of incorruptible beauty, wider than any I had ever seen, its shining bands out of black clouds, a shaft of the spirit plunging into the mountains. In Taos Valley as the sun was setting, I saw fires and smoke in the sky, swirling between the dark slate color of a coming storm that massed over the tops of the guarded, secret peaks. Suddenly a Navajo bluebird of happiness flitted across an open patch of sky that burned with a tropical blue. New Mexico was a magnificent background for the rites of mystery.

I headed for Los Alamos with the idealistic and youthful feeling that if the atomic age was to include me as a participant, then I ought to be allowed in, if only for a look around. It was the old American idea that individuals should have control over their own destinies, and so ought to have more of a say in what was being done in their names. Although a nonscientist, did I not have any rights in the matter? Indeed, should not all life have some say as to where it was being taken? A futile inquiry perhaps, but as a believer in democracy, it was on my mind.

Through my obliging friend, I got a pass for "atom town" and was up at five o'clock in the morning to get the early bus, which was filled with sleepy men and women going up for a day's work. The bus droned slowly up a cliff road past dark pines and cedars as the first light of the sun began to paint the rocky ridges, until, after rounding a sharp curve, it reached a high mesa on top of the world. I looked back where the great, primitive, bare dawn of New Mexico was sending its light over the Rio Grande Valley as far as the long mountains in the distance. Then I was prodded out of dawn by having to get down and show my pass to a "security inspector" at the gate. When they let me in, the dusty streets, the barracks-like nakedness of the buildings, the glaring, institutional whiteness of the town, were such that I might have been back in the army again, trying to make it back into camp by reveille after a night on the town.

I was to meet my friend, a minor cog in the outfit, but it was still too early for him to be up, so I sat in the cafe for a while drinking coffee. When that stimulus ran out, I started to walk around town. I inspected the shopping center and its small, neat buildings: one shoe store; one drug store; one florist; one furniture store; one store with children's toys; one music store; one camera shop; one tailor shop; one beauty shop; one each, appropriately, of what a compact community might need. I also walked through the clean new post office building and out again.

There was an open square between the shops, completely empty at this time of day, with a few benches around it. Feeling a bit uneasy by this time, like a tourist who had strayed behind an off limits sign and could find no exit, I sat down. Then quite a large number of young security troopers in blue uniforms began to walk by. I hid behind a newspaper I had bought at the cafe. Then I was relieved to see that it was time to phone my friend, and he soon joined me; so, pursuing the national habit, we had some coffee together.

That was not quite all I saw of Los Alamos. I visited the high school, where I was told they had some of the brightest kids in the country. Its buildings were also bright, clean, and new. I peered through a glass door into a classroom and saw a teacher discussing theorems with her pupils, something to do with the structure of aluminum.

I inquired at the police station about crime in this made-to-order community and was told that in the past few days there had been a few truck and automobile accidents outside the town, that a house had been damaged by a car, identity unknown, whose tracks were found on the lawn, and that thirteen window shades had been stolen—probably an outside job, since no residents were suspected.

Beyond that, the mysteries were denied me. The public relations officer handed me an information bulletin, officially approved. It had a recurring theme: "We don't know about such things." If a further apocalypse was being prepared in the "tech areas" behind their steel fences, I was given no hint of it. The priests of the mesa were also well guarded, all the way down from the director and the administrative director to the directors and assistant directors of this and that, to some young scientist from a university who might come here for a month or two to do some research and then leave. If they and their secrets were well protected, the logic seemed to be that the rest of us must be, too, and I had accepted enough "security" not to ask too many fool questions that were never going to be answered. That which I did not know was the most of what I carried away with me.

It was discomfiting to be in a place which almost had the status of a religion and to come no closer to revelation than that. If its purpose was a moral one, I certainly got no hint of it. Perhaps moral purpose never came in to a design that was essentially in the service of Experiment. The results came out in the open at Hiroshima and Nagasaki, while the methods that led up to them were now disguised in the name of Security, hardly a religious term.

As my friend and I walked around the town with its too well-placed and overly neat little houses, and its long, white barracks, all with their regulated quantities of use and pleasure, it came to me that the great fact about the place was its complete utilitarianism. Los Alamos was the ultimate in the abstract ideal of practicality, an end product made almost without human contact.

There were no graveyards in Los Alamos, no mortuaries. If you died you were shipped back to the folks in Dubuque or San Rafael. And the residents owned no real estate in those "69,000 acres of scenic wonderland," a plug from the approved handout. A government-created organization known as the Zia Company maintained the town, collected the rent, hired the labor (aside from the administrative personnel and the scientists) and sent the plumber around to fix the leaks. Zia, named after an Indian sun god symbol, and implying ultimate energy, would tell you how many nails you could drive into the wall and how much you might indulge your fancy in planting around your house.

Someone said that there were no friends in Los Alamos, only acquaintances. I supposed that as neighbors there was a little too much they were unable to talk about. Their individual jobs were not free, and without labor unions they had no social control over them. What they seemed to have in common was worry over job security. It was like the army post headquarters where I had been stationed, with many officers busy filing papers in triplicate. Their principal concern, and I confess that I had the prejudices of an enlisted man, was when they were going to get promoted. I sensed a vague antagonism in the air, a hierarchy of people with a nervous awareness of one another.

To be sure, there were a good many clubs and "fraternal organizations" listed as being in Los Alamos, although they seemed too consciously chosen, like the shops, one of each kind. I was told that the Civic Club was particularly active. I talked with one woman about the Soroptimist Club, of which she was one of twenty-three members. I *should* know about it, she told me, I *should* be interested; it was not just of local but of international importance. She showed me their code of ethics, adopted in Kansas City, pledging allegiance to the "sincerity of Friendship; the joy of achievement; the dignity of service; the integrity of professions; the love of country." Every effort had been made to plan an average American town, but I had the feeling that it might be Main Street on the moon. It seemed quite likely that for many residents the disorder of the outside world was much to be preferred.

There was a bland air of unreality about the place, as if it were a perfectly normal thing to be manufacturing weapons that might get completely out of hand and incinerate the earth. Perhaps it was just taken for granted that man, who made the bomb, ought to be able to undo the consequences. As Morris Bradbury, former director of the Los Alamos Laboratory, recently put it: "The whole object of making the weapons is not to kill people but to find time for somebody to find other ways to solve these problems."* Who, or what, that "some-

*New York Times, July 14, 1985.

body" could be is not very clear. To be valid, it might include that whole company of life whose "input" we appear to have been ignoring. They might help remove the isolation we bring upon ourselves by imposing absolutes upon the planet.

I had noticed a fenced in enclosure not far from the shopping center. There was nothing inside but an old hut made of adobe bricks, with a small, locked door on it. I asked about it and learned that it was an Indian sacred ground which had been there when Los Alamos was the site of a ranch school for boys. Through some diplomatic arrangement with the Indians of a nearby pueblo, the little place had been left intact. There was no sign to advertise it, and I was told that a good many residents were not aware of its identity, even though they might pass by it every day.

It seemed to me that this arid patch of ground and its sacred, crumbling hut was probably the one free area left in Los Alamos, attached through earth and spirit to the sky beyond. In religious terms it was a symbol of our human connection with the universe, stemming from a direct relationship with the forces and forms of earth's nature. It meant that in the powers and mysteries of the world, in wind, fire, stars, lightning, and the sun, were the combinations that give rise to life. It was part of the rituals of an even more ancient New Mexico than we were scientifically aware of, going back vast distances in time. Otherwise I felt as if we were symbolically cutting ourselves off from the great spaces of the continent before we had a chance to know them. Was this a turning point in the history of the human race, or would we have nowhere else to turn? It seemed to me that a secret violence without any obligations to life could not even give any importance to death. I took what little comfort I could from the idea that we had hardly begun to know ourselves or the earth which gave rise to us.

I saw a black crow, a free bird, flying over, and decided it was time to catch the next bus and be on my way. I went back to the old New Mexico. I walked toward its mountains, which were matted with snow like sheep's wool. The snow lay glistening on the land. A flock of robins flew low in the pine trees, starting and stopping in the branches. Their red breasts were flushed in the light of late afternoon. They called to each other from their separate points of green, a company of bells. Where the path climbed steeply, I could hear cow bells tinkling down below, and beside me the sound of water running under melting snow, the sound of peace. A little brown rabbit hopped off the path and sat limply under a juniper with its back to me, as if it thought it was hidden from sight; but when I moved slowly toward it, it started to hop reluctantly forward, and then dashed away for dear life. As the sunlight began to wane, a cold evening spread across the hills and caught me by the throat and hands. When I turned back I could see an opalescent sky fading behind the chalky mountains.

Further on, in a dusty town, there was a sign that read: "This is God's Country, don't drive through it like Hell." Driving or driven, how could we help it?

Carl Sandburg, who worked and traveled in the open spaces, wrote:

> The people know what the land knows
> the numbers odd and even of the land
> the slow hot wind of summer and its withering
> or again the crimp of a driving white blizzard*

They knew what the land knew if they were not forced out of it by the removal of all the topsoil to the wind, or the obliterative spread of cities, or the superhighways that sucked the life out of it. Some stayed on, while the sun glared down on

*Carl Sandberg, *The People, Yes* (New York: Harcourt, Brace and Jovanovich, 1936).

the white pavements, and outside of town deer tracks followed dry river beds, and field larks made sudden rushes out of the brush, while doves the shape of teardrops balanced on telegraph wires. They would say: "Seems like too many people are living too fast for life."

Driving westward, I met clumps of gray sheep nosing the bitten grass, a steely sun glistening on their backs. A gray hawk drifted by against the line of distant hills. A cow walked slowly across the highway where the Rockies suddenly loomed ahead like a great wall; and beyond them, down into what was once desert country. the road was strewn with lettuce, "green gold" fallen out of produce trucks, and then I saw the California hills covered all over with houses and grass like green felt. On the fringes of a breakneck city at rush hour, as the cars stopped, started, dodged past each other, and occasionally ran into each other, an angry red sun seeped through a thick haze like blood on a bandage; and the single theme that now appeared to unite the country was restlessness and uncertainty.

CHAPTER 2

Dispossession

The recent changes that have happened to our world coincide, of course, with the vast increase in our power to claim and occupy the earth. This has left "nature," which means whatever we want it to mean at any given moment, in second place, and "wilderness" in arrears. Since the Europeans first assaulted the continent, our attitude toward it has always been self-justifying, even as we left it behind. The Pilgrims and later the Puritans, landing on a narrow strip of shore, said, "God gave it to us." Later settlers and pioneers said they gave it to themselves.

Even before they arrived in the New World, the Puritans had some firm ideas on the subject of property rights. In a letter from England in 1629, John Winthrop wrote: "As for the Natives in New England, they inclose noe Lane, neither have any settled habitation, nor any tame Cattle to improve the land by, and soe have noe other but a Naturall Right to those Countries, soe as if we leave them sufficient for their use, we may lawfully take the rest, there being more than enough for them and us." This right of possession had a longtime precedent in the right of kings, who had claimed authority over all land discovered in their names whose inhabitants were not Christians.

In *Cape Cod*, Thoreau had this to say about an attitude which has prevailed among us ever since we arrived:

> When the committee from Plymouth had purchased the territory of Eastham of the Indians, it was demanded, "who laid claim to Billingsgate?" which was understood to be all that part of the Cape north of what they had purchased. The answer was, there was not any who owned it. "Then," said the committee, "that land is ours." The Indians answered that it was. This was a remarkable assertion and admission. The pilgrims appear to have regarded themselves as Not Any's representatives. Perhaps this was the first instance of that quiet way of speaking for a place not yet occupied, or at least not improved as much as it may be, which their descendants have practiced and are still practicing so extensively. Not Any seems to have been the sole proprietor of all America before the Yankee.

(Note that term improved, still in common use and applied to everything from a thousand acres of bulldozed land to a can of deodorant.)

With that as a precedent, it followed that all living things, even though they had occupied the land from time immemorial, belonged in the same category. They became nonpersons, losing their legitimate rights as continental owners, like the Indians, who were not Indians—we only named them that—and under the new dispensation were only granted such identity as their dispossessors agreed to. If we still refuse to grant any rights to life in nature other than those we define in unilateral terms, then they, and the lands they are a part of, remain in the Not Any class, subject to total improvement. Most of the use and reduction of the land after the Puritans was less a matter of legal rights and moral conviction than of monumental assault and occupation, full of sweat and struggle, roaring and cursing. The open continent was treated as if it were a vacuum to be filled, and at first that was a heady experience.

The great plains were eventually plowed out, leaving only remnants of the original vegetation; forests were destroyed from sea to sea; animals were shot for food or at random; mountainsides were cleared and hills leveled to the ground. "This hamlet," wrote one Samuel Adams Drake in 1880," is the only settlement in the large township of Livermore. Its mission is to ravage and lay waste the adjacent mountains."

Declaring an open season on wildlife was immensely profitable. The pelts of muskrat, beaver, ermine and fisher, wolf and fox, river otter and mink, the hides of elk, deer, mountain sheep, and buffalo were piled high. Boats and wagons were loaded with birds. It was the ultimate in promised lands, full of riotous abundance into the foreseeable future.

Goodbye wilderness. It was turned into a great shadow in the distance, the dimension of a lost appetite. The American Dream rolled over what nurtured it, even expanding beyond the continent to exploit a large part of the resources of the globe. Fueled by unprecedented riches whose exploitation let us go as far as we wanted, we kept moving over to the other side of where we were, migrants on the loose. Finally, we learned how to move around the known world effectively in almost no time flat. This has now become one world that can communicate its messages in an instant, or destroy itself, and all things seem destined to follow the arbitrary nature of human power.

Quick passage made America; it has turned us into a nation of itinerants. How many places do we live in that we can call home? Any and all of them no doubt, but few for long. Interchangeable, or in an economic sense, exchangeable land has become the rule, and so is threatened with a loss of identity. We disposses ourselves. We are still pioneers spoiling for a fight, immigrants destroying the forest, ravagers of soil and wasters of water, never settling down. You might think it would have been enough to equate the hardships and grace of the seasons with our lives; to follow the moon's light along the rooftree; to drink pure water from the well. The tides, calmly, inexorably, moving in and out, might have persuaded us to stay within their frame of reference, but we never had the time. We were always able to escape them. We became the followers of change. As the developer says: "You can't stop change. This place can't stay rural, wooded, and quaint forever." Change seems to be replacing an earlier myth of progress. At the same time, in our preoccupation with motion, we are unable to see the major changes that are happening to us.

On a global scale, conquest of space and triumph over nature go together, resulting in huge areas of emptiness, which does not stop human society from planning even more devastating assaults in the future, although planning may not be the right word for it. The facts about what overpopulation and exploitation are doing to the earth and its species become more apocalyptic every day. This may be all to the good if it produces a new revolution in thought, but have we the time? The inertial process is hard to stop; it proceeds like an avalanche of mud released after heavy rains.

The once "howling" wilderness has now taken on a more respected, even admired character, since we no longer have to tussle with the pristine reality. The term has been switched to describe the human condition, often felt to be a state of disorderly exile. Alaska, or the Antarctic continent, now waiting its political and economic exploitation, may be its last stronghold, but in fact wilderness has almost disappeared in our mind's eye. Taming it has gone about as far as it can go. Taming has come to the end of its tether, having reached a symbolic apex in the invention and employment of nuclear weapons, which carried the process to a vanishing point. It seemed that power over nature could not go much further after the splitting and taming of the atom, although the triumph of abstract thought now turned out to be at the disposition of rampantly unpredictable and dangerous behavior. We have not tamed ourselves.

The split has been widening between us and the life of earth, almost out of inadvertence, as if human society were a force of natural energy gone out of bounds. The results show up in the reactions of the planet. Climates turn erratic because of the loss of forests on an immense scale. Deserts advance; floods spread beyond their normal limits; and with the great increase in world population, now at four billion, natural disasters, such as drought, earthquakes, and volcanic eruptions, claim many more victims, particularly the poor, who have moved into the areas where they occur and cannot easily escape. Millions of people have become destitute in great regions of the world, dispossessed from the starved lands they once depended on and were a part of.

When forests are destroyed, aquifers dry up and soils turn barren, especially in the tropics, since the trees are no longer there to provide their major role in the exchange of moisture with the atmosphere. In Europe, forests are being damaged beyond repair because of acid rain.

"It is quite conceivable," writes Paul Ehrlich, "that if society continues dumping into the atmosphere large quantities of nitrogen oxides and sulfur oxides, which are the principal sources of acid rain, the life-supporting machinery of the planet could be damaged to the extent that humanity will be lucky to survive as small hunter and gatherer groups in less polluted areas of the Southern Hemisphere."*

*Paul Ehrlich, *Defenders*, November–December 1985.

Arid and semi-arid lands around the globe are turning into deserts, at an estimated rate of sixty-seven million acres a year, and this does not exclude the United States, where development and large-scale irrigation projects are drawing out more groundwater than the land is capable of replacing. Large areas of once fertile seas are being made nearly barren of life, deserts in themselves. The population of many species of fish have declined, precipitously. Many thousands of seabirds still die offshore, as a result of oil pollution. The ranks of the once plentiful mammals have thinned out, and plants vital to the existence of innumerable ecosystems are rapidly disappearing. Scientists estimate that species are going into extinction at a rate that is approaching one an hour.

Taming no longer works. On this overwhelmingly gluttonous scale, it leads nowhere but to an outraged planet and should no longer be confused with the advance of civilization, unless we want to outrage that concept as well. The truth is that we have no alternative but to reeducate ourselves in what we came from. Civilized life has never replaced its origins in wilderness, meaning all of nature, the integral life of the earth, and has never risen above our dependence on it. We derive our sustenance from what we can neither improve upon nor finally conquer. The world of life is not our foster child and dependent, it is quite the other way around. There is, finally, nothing but the wild, in its rhythmic timing with the sun and stars; no aspiration lives without it.

Walking recently under high tension lines strung over the bumpy, worn out terrain of an overdeveloped part of Cape Cod, I thought that in spite of all the instant news, entertainment, and conveniences they made possible, all the heartaches and domestic concern they could carry, those wires did not have a very close relationship with the grounds they traveled over. Dolphins communicate with the sea. Our means of communication, on the other hand, become increasingly airborne and detached.

High tension lines seemed to represent an imposition of will. While they helped to unite us, they also symbolized a control that could not let up without extreme risk. The business of imposing human mastery on nature and the planet was a deadly serious game, a win or lose affair. It is as if we were afraid that the dark, deep wilderness we had fought so hard might be let in again. To which any responsible patriot who believes in its reality as he believes in the light of dawn might say: "Why not?" There lies the truth, worth returning to, however long it may take, wherever it may lead.

CHAPTER 3

Wilderness— the Coast of Maine

*I*t is a summer night on the Maine coast. Over water there comes the smell of fish. Television sets sound from the open windows of scattered houses along the shore. Cars nose over the local roads, their headlights probing the dark. There is the droning thunder of a plane crossing hundreds of bays, inlets, and islands as it comes in off global waters, and all around there lies the immense and haunting presence of a wilderness.

While a civilization without foreseeable ends is sending out direct signals across the planet, tightening all the lines through which it circumvents the processes of nature, forest trees of undying ancestry creak and sway in the darkness. In their enduring patience and hardihood, they measure the intervals between one ice age and another, between a summer's burning light and the hissing snow that runs off the pluming branches of the evergreens like waves. Trees are the victims and interpreters of undying weather.

In the shelter of the forest, just beyond the houses and towns, are travelers with bright and darting eyes, communicants out of a major, earth-inspired fear, the bearers of fine-tuned senses through which they find their way and each other. The tide wells up through the islands under the light of the stars, carrying twitching and swirling planktonic organisms, sometimes millions of diatoms and thousands of growing copepods in a square foot of seawater, as well as the shed skins of shrimp and barnacles and the larvae of crabs and other marine animals. Fish fry school and twitch in the shallows, or in deeper waters run ahead of their predators, while on land other hunters explore the sea's edge or roost in the trees. The life of sacrifice, growth's vast potential balanced against an equally vast attrition, but each providing for the other, is part of the earth's dimension. A single codfish may produce several million eggs, of which only a few will hatch out and grow to adulthood. Seeds, pollen grains, and spores are sent out on exploratory missions, fulfilled, more often than not, by only a tiny minority, but in the potential lies the power.

Wilderness enough, many would think, in millions of barnacles covering the rocks, competing for space, squeezing each other out; or in a bed of mussels growing thickly together, the living with the dead; or in the soft shell clams lodged in the mud, siphoning microorganisms out of the water. Sea worms, shrimp, arthopods, and amphipods crawling and flicking under rocks or through the hanging weeds, participate in a nearly mindless probing of their surroundings. What do all those reflex actions, the spontaneous nerve ends, the blind eating and digestion have to do with us? Thinking of all this sum of life as merely reactive makes some people recoil from it and withdraw into higher human functions and attributes. But that extraordinary range of sensitivity is as vital as the sea that spawns it. We can never completely detach ourselves from it. Out there is the great night of being.

On the fringe of these teeming worlds, hearing the cold currents of the tide run by and its waters trickling through the rockweed, lifted by the exalted cover of uncountable stars, I have felt a whole inclusion that I have never experienced anywhere else. If God is dead, or missing in action, this might be the place to recover him again.

When I travel out in a skiff, a fairly slow boat that will not deprive me, through its speed, of the sense of what I meet, I know that these shores are full of dangers that are nevertheless rules for the living. The beauty of this electric, silvery gray atmosphere is allied to grim necessity. I could not survive there long, left to my own devices. I guess my own world is largely out of practice. The shores and islands are full of rocky crags, hidden ravines, and unexpected pitfalls. They combine great allowances and at the same time dire accidents for the unwary. The polypody fern in its crevice in the rock, the porcupine hitching slowly up a tree, the osprey on its high nest in a dead spruce, know this region better than I do, who have been too long removed from it. To really live, as opposed to residing, there, I would have to divest myself of any number of contemporary substitutes for reality. Yet as it is our original home, I have to peer deeper in order to advance my sense of earth location and my hidden needs. Communion with life means understanding its skills; otherwise it becomes a junkyard of castoff opportunities.

The seals that lie out on rocky islets in the sun during times of high water and explore the channels where the fish are know these waters down to the finest sensitivity in their whiskers. Their bodies can feel the currents in all their swirling changes. Their brown eyes perceive an intricate marine distance of innumerable facets and corridors stretching way beyond the human capacity to see; and while the seals explore these waters, young eiders, all through the summer months, are being led by adult females through miles of coves and inlets, an inward training in their future migratory headings and in their home range.

The wilderness, once feared for its darkness, then cut down and made "livable," can now be bypassed. It is outside of us, where we always meant it to be. The wild and uncultivated is no longer needed in a world that only looks out the window at itself. Yet every life still obeys its laws. Wilderness is the most gloriously lighted space for existence on this earth, which is something of a contradiction in terms, since it is the earth, its true measure and our foundation.

The quality of the wilderness needed the wolf, now in exile, and it saw through the eyes of the lynx, an endangered species, whose name is derived from the Greek *lugx*, or light, because of the intensity with which those lucent, yellow globes looked out upon their world. Among the Pawnees of the West, the bobcat, now very scarce in the populated East, represented the stars in the sky, because of the spots in its furry coat. So did the fawn, with its dapples of white on a dark background. When native people lived with what they could never call wilderness in our sense of the term, they shared it with the rest of life. It was filled with animate powers, benevolent or fearful, but always inseparable from their feelings and inherited beliefs. Celestial motion and the continual motion of the weather were allied. The land led to the sky.

The voices of forest and shore, along with the clicking and sighing trees being moved by the wind, still include the hoarse and gravelly tones of the great horned owl hooting in the shadows, while a startled great blue heron which has been roosting in a white pine tree leaves it with a strangled cry. Between the hard, studded limbs of living and fallen trees, there is aisle after aisle of briars and hay-scented fern, a broken sequence of

feathery, shaggy and splintered surfaces, while the wild, cel-
lular nerves of root tips explore the ground beneath.

This calm, original state, yawning deeply out of the past,
with the stars, its ancient friends shining and guiding in, is
surrounded and invaded by creatures who only half live there,
since they pride themselves on not having to live on that daring
edge of subsistence. The dark ground with its night-conscious
trees has become unfamiliar to us. We seem to have lost the
speech we need to converse with it. Still, it moves us, as if the
light on the branches under the vaulting night sky, and out on
the shining tides, were speaking to us with a more compelling
voice than any we have heard in a long time. The wind shoves
at the silvery waters and rises through the trees with a certain
equanimity, a long enduring flattery; and perhaps the plants
and animals are listening too, waiting for the day when they
can freely move again, because the monumental scale, in life,
in death, will be open to them.

CHAPTER 4

Fire on the Mountain

I live on an increasingly crowded shore, where the house lots and buildings are so overpriced as to make you think all Americans must be millionaires. Few of us are as justified as a hunter or a fisherman in claiming kinship with the life of land and sea that surround us. Most of us are culturally removed from a dependence on where we live, and in a sense this results in an emptiness in the environment itself. Not that we are conscious pillagers and plunderers necessarily, but if there are no fishermen, there are no fish, as without the clammers, for all their continued brutalizing of the sandy flats and muddy shores, there would be no clams. If a bird does not illuminate our sight it falls prey to our own blindness. The substance of the living world begins to disappear. We do not know what we are unable to see, or go out to see. Hunters and explorers have to see themselves in what they hunt in order to bring it into their persons. By comparison, we are turning into viewers and curiosity seekers, only entering the waters when they are warm enough. It amounts to an uneasy kind of possession through which we bring in values from the cities, the superhighways and superplanes, technological advances and economic risk, a a world of summits and devastating falls, all to some degree detached from the continuity of the tides.

Unconsciously, we still count on some half-expressed, subterranean foundation in nature to sustain us, no matter what we do, and for that reason are decidedly uneasy about our ability to destroy our links with it. It is as if a new need to see, to let outer life in, had accelerated in proportion to the speed with which our lives and experience are allowed to pass by. The human world has been flung out, sent everywhere on its own errands, and in the process the earth is being opened up to mere chance. Can this uprooted, and uprooting nature of ours find a place to settle in, where we can recognize ourselves in the measure of living things as they go on about their endless rediscoveries of the earth, following the primal terms of light? They are the originals on a lasting frontier that we have hardly begun to take in, although its distance is embodied in everything around us.

Halley's comet came into view recently. It was in the news for weeks and could be seen on nights when the sky was not too overcast. I saw it through the telescope of a college observatory; it was dead ahead, low on the horizon, a fuzzy but intense light coming at me from seventy million miles away. At the same time, I felt a whole sky rolling over that small, limited opening, filled with great stars like bioluminescent creatures in the ocean, an endless flying, a coming in of the glorious outwardness of procreant space. In primal fire and light is the only unity.

I remember a fellow draftee, in the army camp where we were in training for the infantry, who, in his Texas accent, used to yell out: "Fire on the mountain!" every now and then. In my mind's eye, I saw the people of whatever small town he came from as they ran out to cast their eyes upon the neighboring hills. There are two kinds of fire, that which creates and that which destroys. The world of life in nature employs them both. Whereas we are caught between the two in a precarious state of our own devising, the combined fire in nature is behind regeneration. It is in the foundation of mountains and the state of the trees.

One New Hampshire mountain of moderate size, called Mt. Cardigan, which I have been climbing since I was a boy, has a bald, rocky top, the result, I have heard, of a fire, which may have been accidental or set. The farmers once ringed Mt. Monadnock so as to destroy the remaining wolves. During the month of October you can gaze off from this summit at hundreds of square miles of trees flaming with color across a waving, billowing landscape, accented by the dark green spires of spruce and fir. They climb in waves of embattled growth, the sick and healthy, young and old, living and dying, while the great autumn shadows shift and plunge across them, conspiratorial and protective.

From the lower slopes to where you can only keep going by climbing the sky, the elevations are full of of changing spatial relationships. White pines and relatively sheltered maples and birch lead to tundra plants exposed to the worst the atmosphere can bring them. You leave behind the abandoned pastures long since grown to trees, as well as the cellar holes and occasional family graveyards whose stone markers show the evidence of abandonment. Here whole farm communities once dug out the rocks, plowed the fields, built houses, barns, churches, and schools, only to move on because of world forces beyond their control. The whole continent is now occupied by millions of their descendants, who seem to have forgotten what they left behind. Every life is migratory, but it is as if we were the one species which had been experimenting with a disengagement from the nest, or living space, that other animals inhabit, or to which they annually return. I imagine we think it is a luxury not to be so confined, but inside us all there is a darker home that lies in waiting, timed to tides and trees, a place of recognition.

One day, I heard a tiny, whispered call and located two pine grosbeaks in a spruce tree, where they were leaning down as they pecked at its boughs. The females had nut-brown heads, gray bodies with touches of copper and green, while the plumage of the male was a suffusion of pink and red; it was a peach-colored bird, with dark wings. They were accented to perfection against the dark needles of the tree.

On a bright day in the mountains, spruce needles shine like glass, and the air slides through them with a sound that is hard to distinguish from water. But a cold, clear brook thunders by, running over brown and gray boulders, a sounding out of some original silence. It whirls and rushes down toward some neverending, and, like the singing of a birch tree in the wind, it is a true voice, the right one to listen to.

A light wind lopes along, an ambulatory wave through the trees, and as I walk along it brings me a sense of the swirling, conflicting powers of global weather. I am lifted by storms, mixing, forming, and coming on, to an extent I could never have imagined by following the weather reports, derived from instrumental order. I am an instrument of unpredictable motion.

In November, on the higher elevations, the trail, full of boulders and dead branches, and on its level spots very muddy under fallen leaves, begins to turn ice hard. Clusters of elongated crystals like teeth have thrust up out of the mud. The surrounding environment has changed its mood and character in recent weeks, as if the mountain, in response to the gradual advance of arctic powers, was inwardly preparing and resettling itself. This new, stripped composure is part of the timeless turning in the universe and might have to endure for a long time. Who knows how long. On the tips of stunted spruce and balsam are tiny beads of ice, and rocky crevices are lightly sprinkled with grains of snow.

A blue-black raven soars off beyond the open ledges, crying out with a gutteral "waugh!" It rides with an easy, silky style over the slopes and shoals of air like an otter on its slide. One morning in early October, I watched a red-tailed hawk doing some fancy maneuvering in the wind, wheeling down at a long angle and then sweeping back up again, and the hawk, evidently a young one, full of mettle and adventure, plummeted down on top of it, while the raven contemptuously slid off to the side. The redtail then paused, legs and talons extended, and made another drop for the sheer pleasure of it. Having been fired up by the raven, it then tried a few more short practice drops before flying out of sight. Here were two species allied in the adventure of a continent that had been evolving for hundreds of millions of years. As I watched them it was not how distinct they were, one from the other, that impressed me, but the freedom in their distinctness.

Ravens seem like almost phantom individuals, with a powerful ancestry. They tease me, in my rare meetings with them, toward recognition of a speech we have hardly begun to translate. I see ravens as my progenitors in spirit, birds of ill omen or good luck, all in one, of dark, mischievous minds, and of a sardonic pride. One flew quite close to me one day as I was standing next to a spruce tree on the Maine coast. It swished past me on stiff and silky feathers, carrying a mouse in its bill. I sensed enchantment in its character. The field guides fail to mention that. Perhaps we are afraid to stray too far beyond accepted nomenclature and the scientifically proven. We may sense too much of our suppressed anarchy to trust in merely intuitive equations, with their essences and vague parallels, but with the will to be haunted, I am leaving an open space for ravens. In any case, they got here first, and if they stand in with the sun and moon on intimate terms, as the Indians saw it, then they have lasting powers I ought at least to respect.

In this crowded world, our sense of coexistence with wilderness-life can be enforced by heights that are hard to climb. They require at least some equation in us with areas of life on which we have not yet slapped a lien or a foreclosure. That goes for mountains too high to be appropriated as well as the sea, of whose depths I know next to nothing, except for what I have found washed up along the shore. Mt. Katahdin, which was saved from being hidden by development or half destroyed by that environmental degradation that is becoming a global curse, is one of those mountains that remains in charge of greater realities. For that, all praise and honor to Percival Baxter, who purchased the land and gave it to the state of Maine, to be kept perpetually as a wilderness area. He must have revered its existence.

The last time I climbed Mt. Katahdin, I was told by an awestruck young camper coming down the trail in the morning that he had seen Baxter Peak being struck by lightning the night before. Mountaintops, for the religiously inclined, or even for pragmatic innocents, might still be thought of as the seat of the gods, playing arbitrary games for our confusion, while at intervals Jovian bolts are hurled down on the bare rock without equivocation.

Katahdin stands off by itself above lands around it marked by moose tracks and covered with lakes, rivers, bogs, streams, and beaver ponds, while beyond its influence the roads and highways radiate out in all directions to serve our world of goods and services. It dominates the world around it. As I hiked in closer and the mountain rose in front of me, I felt as if I had come into the presence and was prompted to take my hat off. Its great walls, its bowls and shoulders, invite the roaming wind—it is also a creator of winds—and it continually exudes the rich, hard messages of the life it shelters.

The woodland plants along the trail, wood sorrel, gold-thread, bunchberry, clintonia, and many others, climb under the protection of the trees until the cover gives out and the soil becomes too thin for them, or nonexistent, and there the low, alpine plants take hold. I caught sight of a red-backed vole as it darted across the trail and disappeared into the rocks. This was a character who was little known to me, and the fact that it was an intimate of the mountain in all its stress and storm seemed to give it an added stature, although I suppose that mice, having a pivotal role in survival, are in no need of our compliments.

Later on, on our way down from the summit, my fellow hiker, Fred Turner, and I walked up hesitantly to an enormous bull moose lying in a thicket, which is certainly not the way we ever expected to see one. Its antlers showed up like the top of a big pile of driftwood. The great mahogany animal was colossally indifferent to our presence, as it lay there with no motion except a slight twitching of its ears.

The Katahdin spirit is full of incorruptible danger. The great weather never relieves it, bathing it with sweats and frosts, covering it with dense fog, lashing it with icy hail and violent winds. The mountain takes all that comes to it, as some patient, omnipotent being would; and the plants and lichens that grow over the rocky tableland at the summit cower and spread wherever a rock face, crevice, side stream, or snowbank gives them some purchase or protection. After the ice sheet retreated, they were the pioneers that gained a foothold on a then bare and sterile mountain, to expand and elaborate on their communities as time went on.

The plants are on a frontier of life, coping with conditions beyond our mammalian reach. They face intense cold and winds that may attain a velocity of a hundred miles an hour. Sleet and rime ice cover them. Although they get enough moisture to survive from rain and fog, they could be quickly dried out by the wind. They are subjected both to intense ultraviolet radiation and to low light levels, since the mountain is covered by clouds for much of the year. Their tenacity is astonishing, although a great many explanations are available for the devices that make it possible for them to persist, from a pigment anthocyanin, in their leaves that is capable of converting light into heat, to waxy leaves that reduce dessication, or that have a layer of hairs that reduces evaporation by cutting down on wind velocity. They also grow low to the ground so as to modify wind damage. They are very slow and deliberate in their growth and in their respiration and spare in their flowering.

Plants cannot be accused of poverty because they grow in poor soil and bitter weather; they are rich in collaboration with it. Not only can they add to science, as if that were an end in itself, but they are a dimension of life that challenges the mind. It is their complex affinity with extremes that makes you gasp in thin air. They are universalists. Plants have tried everything, in every possible corner of the earth, under every condition, and in a still unknown variety of ways. They seem to reach as mountains do, after eternal principles. We can surely do no better, whether by making them serve our purposes or through the pretense that we can surpass them with our technical solutions. It is they who have the foothold. Their growing extinction in many parts of our world only implies that we are losing our own hold on aspiration and restraint.

Up to where the alpine communities bloom *in extremis*, monolithic granite blocks, speckled black and green, are piled up in slides across the flanks of the mountain and lie across the trail. One immense, square rock stands out like the displaced plaything of some divine giant born of geologic time. He might have shaken it out of the greater mass of the mountain, then left it there, part of the offhand, thunder and lightning style of building he was engaged in.

On the summit, which may be made more refreshing after the climb by eating an orange, or reading Persian verse—as an Iranian companion did on a previous trip I took with my daughter Kitty—you are only a visitor. You cannot stay up there too long, or the night might lose you, or the mountain standards become too much for human frailty, if not for those plants with their foothold on some eternal summitry. If they are walked on too much, they may be severely damaged; but I think they are not as fragile as are we, who have protected ourselves against exposure to the extent that we begin to find the elemental messages almost too faint to hear. The mountain is too much for us; we could not live there. At the same time, it invites you out on a wing of the universe, as you rest on Katahdin's "Knife Edge," to experience how limitations may lead to the unlimited. So, as I hobbled and shambled down those rocks into the roaring brooks and sundown-shaded valleys, I left the mountain behind me, under a mountainous sky, to decide what things were worth.

CHAPTER 5

Earth's Eye

I have a color print that a friend made for me, from a close-up he took of the eye of a sharp-shinned hawk. It stares me down. I am aware of a few of the facts about the binocular vision of hawks and other birds, or I know where to look for them if I need to. I have been told that the eyes of a great horned owl have one hundred times our ability to see in dim light. A sharp-shin can spot a mouse in a field from a thousand feet up, having eight times our acuity of vision. The varying ability to see has a great deal to do with the requirements of any given animal in its habitat, down to those in the ocean depths, or in caves, which are not able to see at all. But it is not the factual detail that counts so much as the extent of these earth-sprung visions themselves and their collateral responses.

This sharp-shin's eye is green, the color of a young bird; it turns red in the adult. The head feathers rake around it like streaking currents surrounding a whirlpool. The round black pupil at the center of the green iris threaded with its network of tiny veins looks like the earth's shadow, a black cap moving over the sun during an eclipse. Perhaps, in its central brightness, it might remind you of the eye of a hurricane, passing by overhead in the great cover of inexorably moving clouds. Green as grass, red as fire, the eyes of hawks have been seeing into the land of America for longer than we know.

In the 1830s, George Catlin, lying in the grass and looking out over the rolling prairie, felt lifted "upon an imaginary pair of wings, which easily raised and held me, floating in the open air." The high grass of the original prairie, dominated by the big bluestem, which grew to seven or eight feet tall and had a deep and extensive root system, would periodically catch on fire, which would spread with awe-inspiring speed, easily overtaking a man and his horse riding at full gallop. In his book, *The North American Indian*, Catlin describes the prairies on fire as "one of the most beautiful scenes to be witnessed in this country." The Indians believed in the spirit of fire and sensed its coming, mixing "all the elements of death." When they gazed into the howling wind, the thunder and lightning, the immense clouds of smoke extending over the vast plains with their beds of liquid fire, they encountered it in their souls. The fire spirit was inherent in the majesty of space.

The hawk's eye not only picks out its prey from great heights, it also sees into it and what surrounds it. So it is a knowing eye. The earth's particulars are in its mind. We are addicted to proving phenomena, picking out single objects, facts, and images from their context, so as to help us draw conclusions. To assess an entire landscape constructively means to be able to think in ecological terms, but this does not necessarily imply that we know it well enough to see *in*to it. The hawk may be restricted to its kind of environment and its specialized means of dealing with it, but it also acts and sees in terms of its constant variants and complexities, and not from the outside in. It is part of a system that must have mystery added to it in order to match the fiery phenomenon of the spirit. The answers the hawk might have to what and how it sees lie in the spontaneity of the earth itself, which it follows to the end.

Then there are fish, circling in the illumination and deeper shadows of the sea, schooling ahead on oceanic migrations, their lidless eyes staring ahead, or lying silent over dark stones. They are ever conscious of each other and what lies around them. Dragonflies, with their absurdly large, compound eyes, dart stiffly over the surface of a pond; a butterfly drops down into the cover of grasses when the sky grows overcast and flies up when the sun comes out. I watch the sun lighting up a flock of white shore birds as they speed off, like a swinging basket in the air above the sea. They can sight a hawk from a beach and start looking up and engaging in nervous movements when it is only a speck in the sky, minutes before I can spot it, even with a pair of good field glasses. These are just a small part of the great company of visual acuity and sensitive response. They are not just single, separate individuals divided in their abilities but mutually engaged with the life around them. They dance with the light, the wind, the mercurial motion of the water, the waving land.

Not long ago I watched while an individual of the amazing race of squid, one of the cephalopods, related to the octopus, was being put in a salt water tank by itself. It immediately started to react in an irrational, disturbed way by darting around very quickly, expending its energy, squirting a cloud of ink, and then dropped down to the bottom. This was not so much a matter of neurotic behavior as we understand it but a reaction to isolation and a fear of predators. With a little sympathy you could feel that terrible shrinking inside yourself. Squids, like most other marine animals, live a perilous life, one that continually anticipates escape, and they become most vulnerable when individuals are caught away from the crowd. One of the functions of the ink is to desensitize the chemical receptors in the fish that hunt them, since fish rely heavily on their sense of smell.

Squid are also equipped with extremely complex and efficient eyes. "Wonderful eyes," Alistair Hardy called them in *The Open Sea;* and these eyes can transmit immediate messages to the pigment cells in their skin, so that squid can change color with astonishing rapidity. Their color can switch from red to white to black and back again in any combination. At times they look opalescent, pink, bluish, or pearly gray. They are also able to alter their skin patterns by adding streaks, bars, or stripes on those tubelike bodies with their rippling fins. By this means they can hide when floating in seaweed, or even mimic other creatures that are close by. They also add them when expressing their feelings, when hostile, friendly, or in courtship. "Blushing in streaks and bars" is the way one scientist put it. A squid is a highly mobile animal, with a maneuverable and quickly accelerating body, and with those changes in tone and color it can express its feelings more openly and frequently than most birds or mammals. This changing in pattern and appearance is as elaborate, perhaps more so, as all the posturing and motion of birds that goes with their singing; it also implies a highly developed self-awareness. They may not be reflective, but they are intelligent; they understand the signals of other species, and the way they communicate their moods is astonishing. Blushing goes a long way.

Do we not court each other, or express our hostility? As performers do we not change color, or draw lines, streaks, or stripes on our bodies, and when too long isolated from our fellow men and women do we not become neurotically inclined? Anthropomorphic allusions ought not to bother us so much that we hesitate to draw parallels with other lives and, above all, extend our sympathy toward them. If only because we have put it, and all other wild animals, at such a distance from us, I would welcome an exchange of sympathetic reactions with a squid. My senses are in need of rescue.

The hawk eye is a little disturbing. It does not look into mine, as we expect people to do when they are asking for our trust or understanding, or simply making some comment about burning the toast; it looks through me to a world beyond. It could not care less, obviously, about my reactions, or the preoccupations of mankind, even as fellow predators. If a sharp-shin knew us that well, it could only have contempt for our blind wasting and bludgeoning of an earth from which we are supposed to have attained our skills. The hawk's eye looks past me with a primal intensity, out of a seating in the elements, the eclipse, the whirlpool, the stations of light, the profound equanimity of a wilderness from which it gains its pride, from which we withdraw to become more psychotic than a squid could possibly be.

Putting nature down in favor of our own "superiority" over the rest of life is our stock in trade, but in so doing we not only excel in the rigidity of our thinking, we begin to lose our ability to see. The earth is full of other travelers, billions of them, a visionary fellowship we are very much in need of. Perhaps we become most aware of that when we find ourselves with nothing but ignorant and detached machinery to give us our directions. In other words, we lose the anchor of vision to the mysterious currents of the earth, shared by all the world of life.

Why do baby turtles sprint for the sea after they are hatched? They have never been there before. Why do young sea birds fly halfway around the globe to places they have never seen? What is there in the head of a migratory fish like an alewife that makes it return to the same stream it grew up in after three to four years' absence? Many theories have been brought to play on the mysteries of migration and the mechanisms that might be involved, and none of them quite hit the mark.

Tiny chicks that have never seen a hawk before will dart for cover when one flies over, but they do not react in the same way to a less dangerous bird like a pigeon. The hawk, say the scientists, is a sign stimulus that sets off a nervous impulse buried in their systems. This is not, as Joseph Campbell pointed out in *The Masks of God*, a matter of individual reaction. "The responding subject," as he put it, "is, rather, some sort of trans, or super individual, inhabiting the living creatures. Let us not speculate here about the metaphysics of this mystery; for, as Schopenhauer sagely remarks in his paper on the "Will in Nature," 'We are sunk in a sea of riddles and inscrutables, knowing and understanding neither what is around us nor ourselves.' "*

The sea image sleeps within the tiny turtle so that it heads for it as soon as it wriggles out of the sand, and the chick knows the image of a hawk, as all chicks have before it. The earth calls forth an endless array of spontaneous responses, out of a depth of inherited vision. The reasons for it may be impossible to solve: but if they lie in the inscrutable they are also manifest in the life of a planet that in reality knows its distances and acts on them. How should I know where to go, how to respond, even what to say, but in their company, which is a cosmos of multiple vision? They travel past me, but guide me at the same time.

The earth changes in spite of us, the way the watery atmosphere is changing even as I write, felt in the trees, understood in the waters, followed by birds. It is our medium. There is no substitute for the eternal weather.

A magnificent autumn wind, expressing a major exchange between warm fronts and cold, stirs the tall white pines into

* Joseph Campbell, *The Masks of God 1: Primitive Mythology* (New York: Viking, 1959).

powerful expression. Timber-loaded, sinewy trunks wave in the wind, the upward curving, sweeping branches giving and dancing, nobly bending and bowing down, springing gracefully back again. Thus majesty leads to majesty. Some image also sleeps in trees. Do they not have a secret receptiveness, even if they cannot be said to see in an outright sense? Are they not a "seeing" response to the wind, this growing darkness and this waning light, as they are reflected in the motion and fiber of their being? While the wind roars, I put my ear to the trunk of a pine and hear the air hissing up and down, fairly crawling over that great column with its skin of rough bark, and it sounds like rushing water. It is part of an inheritance that we have been long neglecting, and to our peril.

CHAPTER 6

A Link with the Sea

There is an inwardness afflicting the contemporary world that is not caused by the pursuit of self-knowledge so much as by forced containment. We are in need of those outer reaches we left behind. Having once had it pointed out to me, I used to walk out across the sandy flats at low tide to look for the Pandora clam, a tiny, pearly shellfish shaped like a delicate earring, almost too thin and flattened to hold a functioning body. That was the furthest I could go and still meet such strange and exquisite marine creations, and hold them in my hand. Beyond that . . . a great suspension and inclusion, something beyond drowning. In common things are greater extensions of ourselves than we ever conceived of. We may think that as a race we travel beyond all limits, but we have hardly started out in life.

The gulls parade over the wet sands at low tide like paradigms of seasonal change. The wind and the stretching waters are their responsive guardians. Beyond the long line of shoals in the distance, the waters turn green with a new growth of plankton, while shallower depths inshore are still golden and icy blue. Under the light of the sun streaming down through shifting clouds, water and sand beckon away with their continual motion, and the birds follow with their shivering, whistling cries. Spring mystery is on its way, the inevitable, still unraveled, coming on.

It is a greater power that invades us, in all its beauty, at first a silvery air hanging between gray trees. Snow showers move in with thunder and lightning. Squalls darken the sea, while the grasses grow greener. There is an offcenteredness in the change which the almanacs and the calendars are unable to measure. The spring flies dance like tiny shadows in the fresh light that bathes the ground, and the soil stirs with each of its wakened communities, responding in its own time to the universal timing. The waters of lakes, ponds, streams and rivers, marshes and estuaries, run and swirl ahead toward the ocean, carrying the fruits of the mystery. Alewives, coming in to spawn, circle at the mouths of outgoing channels where they meet salt water, all the way from Canada to the Chesapeake.

If you watch and wait long enough by any shore, the messengers of life appear from all points of the compass to tell you where you are. So fingerling eels, spawned in the mid-Atlantic, wiggle in to local waters; and shad, salmon, and alewives, or "freshwater herring," approach river mouths and estuarial systems up and down the coastline. These and innumerable other migrants are part of the interalliance of the continents. As their numbers dwindle, and as we ourselves become more crowded and detached, so does our sense of a planetary freedom of motion, and that tension with the elements felt by living things as they circle on to find their way.

When I first saw the massed assembly of alewives in the

local brook where it ran under the road, they seemed to bring something to me that I had been lacking. It was an intrusion of energy that was not beholden to anything the human world could do. I found out as much as I could about them from the biologists and the fisheries people. I learned about their swim bladders, lungs, livers, lateral lines, reaction to light (photo-taxis), the function of their nervous and digestive systems. All this and more helped me to see what kind of animals they were, insofar as you can ever see into the enigmatic being of a fish. I sensed, at the same time, that what really drew me toward them, as they themselves were drawn in out of the ocean in the spring like iron filings to a magnet, was their embodiment of a wider range than any I had ever entered before, or been encouraged to enter. Following them, up and down the length of the stream, began to open up the waters for me, a link with the life beyond my self. It started me on a migration of my own, a process that cannot be completed in anyone's lifetime but is part of the passage of all global communities.

These eleven-inch, silvery strangers also helped rescue me from outdated attitudes that saw them as "nothing but," or "only," fish. Of course, there is a time in your early life when you sense that this is not true. You are what you meet and are able to receive. The fish, although strange, join you. They enter into your own novelty and strangeness, your terra incognita, often to be registered there for a long time without a name. But we are educated to civilize things, which often means to dismiss them.

Where they moved in from salt water on an outgoing tide, they swam through a sinuously curving channel that led through the salt marsh. Then, in separate groups, they worked their way upstream toward brook waters pouring out of the ponds, where they would spawn and then return to salt water. As they kept traveling up the narrow valley, steep banks on both sides, there were lengths of stream where they struck the waters like a shock wave. They changed the silvery marine color of their scales to pinky yellow and light tan to match the colors of the stream bed. Their big, lidless eyes stared ahead, fixed like mirrors of the sea, a pond, or a drop of water, open to the light of the world. Can fish be said to have souls? Carl Jung wrote that the soul is an "eye destined to behold the light," requiring "an unfathomable depth of vision." Out of the psychic depths of the waters that created them, fish must qualify.

They swung up against the downflowing current, circled in midstream, or skittered and dashed across. They fled toward the shaded banks to avoid hundreds of gulls shouting their hunting cries overhead, and at the head of the valley, where they were crowded into concrete fish ladders built to help them rest, as well as to surmount a torrent of water flowing over the rocks, they met the human race. A hundred yards upstream and they could slip, one by one, into the ponds where they would circle and thrash in the shallows, spending their milt and roe, but there they were dramatically detained.

Mobs of children yelled and screamed as they plunged their arms into the water and tried to catch the slippery animals. Men and women joined them with nets, or stood watching from the bank. At one time, when there was a much smaller population of townspeople who valued these fish because they depended upon them, the annual migration of "herrin' " was an event to look forward to. It meant added bounty from the sea. Now we look into the water from a passing distance, and the annual run becomes a spectator sport, except for those sports fishermen from out of town who can make money from a free source of bait and so demand a greater share of the produce, in spite of local regulations. The people and the cars come in from everywhere, but the real need is gone, although the event brings its lasting excitement, with a hint of anarchy on the fringes.

Every year the run is a center for a wild meeting, one world of air, scarcely known to itself, and another one of water, moving up with marine authority. Some children try to help the fish along where they are milling in the ladders by throwing them a little farther upstream. Or they do just the opposite, throwing the frantically flipping, gasping bodies out on the banks. Contemplative people have often wondered what compels the fish to expose themselves to so much risk and sacrifice, although that fire can never be avoided. They often put it down to their not knowing any better. This migration has been attributed to the "blind drive to reproduce," which I have always thought of as a demeaning phrase. We would not like to have it said of us. I suppose we question that exposure because we are afraid of its equivalent in ourselves. Enough uncertain outcomes are involved in trying to run things our own way.

The alewives dying on the grass are pure white at the keel, shading up into colors that arc of a sea metal, a silver shifting to pewter, with a sheen of rusty copper and gold on their backs. Why should we instinctively plunder anything so beautiful? Perhaps because the Lord throws down beauty like a shield, and we can only accept the challenge. Wanton, playful, predatory behavior is as much a part of us as conscious thought. We eat, we devour, therefore we are.

They shake, shiver, and shimmer like water falling down a mountainside, in rainbow light. Each is like a stout, animate leaf full of coursing veins. When you cut and clean one, the blood and mucus stick to your hands, the great common material, the stuff of which the worlds of life are made.

There is everything in this annual event of injury and protest, of inescapable hunger and need. Alewives brought me to the no-quarter part of the cosmos. I felt for them as I have felt myself tossed between bottomless pits and unattainable heights. We do not fit. We hang on like the fish for dear life. Underlying earth motion and desire give us no rest.

As I have watched them migrating back to salt water, wearily letting the current carry them downstream, they have suggested all kinds of unfinished qualities and connections. They will come back, but there is never enough time to know them. Why should I stop at our incomplete ideas about the awareness and intelligence of fish, for example? How, in my ignorance, can I reject them as being too limited to teach me more than I already know? Once again, one of the many worlds of only partly known currents and deep affinities retreats beyond my sight, but it draws me as if it knew me.

The marine horizons contain the ends and the beginnings of life. Even the fateful entrances and departures in human history have been swallowed up by those restless waves and inexorable tides. Modern nations have fished the oceans for all they were worth, staking their own livelihood on the outcome. Fish populations have plummeted, although, given enough time and relief from pressure, they might regenerate. The resources of the sea are finite, although the sea converses with the infinite. We cast our nets out everywhere, until the supply is gone. Then we move on, taking more from the planet than it is able to give back, dealing as usual in one-way risks. The original source can give us no reassurances, which may be what its silver messengers, the inland spawners, have to tell us, year after year. Their silent casualties offer no help, but their returning, circling ahead like the globe itself, their daring in the face of everlasting odds, is as common to us as the struggle in our hearts.

CHAPTER 7

The Real World

*I*f we think plants and animals only pursue some lesser and limited function we can easily name, and, for the most part, put down, we have little basis for a belief that they play a major role in our lives. Outside of our own power-minded and staggering affairs, they appear to go about their business in ways that are known to science but are principally a source of curiosity to the rest of us.

For a great many people, trees can't communicate and animals can't think. We judge them according to their ability to be useful to us, or even to be like us in peripheral ways, but we put them in a world outside. This denies the saving reality that the whole genius of the earth is to contain multitudes of distinct communities, each of which expresses the profound capacities of life in its own way. The brilliant variety of their perception, sensitivity, and quickness, endlessly exchanging with wilderness energies, escapes applied knowledge.

I have been watching some black-capped chickadees at the bird feeder outside our kitchen window. They flit back and forth, landing on it, picking up seed and flying off, giving way to each other as they come and go, with occasional displays of aggression that never last more than a second or two. This need to give way in the interests of order and space is universal. It is even characteristic of planes landing or waiting in line at an airfield, but the little birds respond to deeper ties. As they give way to each other, they reflect a courtesy in nature, and they act in accordance with an unconscious sea that plays no favorites.

Through science, we know much more about animal communities than we used to, which does not necessarily imply that we credit them with sharing existence with us. With some justice, we do not see them in our depths of misery and heights of joy, in the human sense of a whole universe of mind, of tumult, conflict, and unknown voyages. And yet, as they begin to be turned into displaced and missing persons, we watch the earth being diminished accordingly and our voyages cut down. We need their company, so that we can have other eyes to see with, other ways to navigate, other senses to point the way to what we have not yet learned about the earth's beauty and capacity, and about ourselves.

I once had the idea that all birds appeared out of nowhere in the springtime, like the robins, made their nests and laid their eggs, raised a family, and then disappeared into nowhere again. It hardly crossed my mind that the pigeons cooing in the city park and the robins in the country were mutually engaged in a planetary exercise. Many years later, when I had moved to the seashore, I watched the terns that migrated in during the month of May, to leave in mid- to late summer and return the following year. They led me on, and I began to see them as forming a vital link between the places where they nested and the coastal seas beyond. For them to choose a particular stretch of beach, an offshore island, or a sand dune to nest in was to distinguish it. It takes an exceptional form of life to know the grounds for life.

In the electric spring light, they skimmed in off salt water to start occupying a sandy island on the far edge of a salt marsh, just behind the beach. With their bouyant wing beat, they would swing down the inlet to the marsh and hover over the water, to dive into the surface like little arrows and pick up fish. Their voices were harsh, vibrant, and shrill and complemented the sparkling sands and the waters struck with light.

(The terns are a race whose local nesters are being pressured to death, both by our wholesale occupation and by the gulls, whose numbers have greatly increased since the turn of the century. The gulls are opportunistic scavengers who take advantage of the waste foods a spreading civilization provides, and they are crowding the terns out of their chosen habitats. The terns were almost obliterated a century ago by being shot off for the millinery trade. With protection, their numbers began to climb again, until the developing twentieth century whittled them down. Year by year, their population declines. A great migrant, a striking and vital part of the life of the shore, is becoming a marginal species.)

Whatever I could learn from watching terns on my own was not quite enough. I needed interpreters, and to see man and terns together. That is why I started to visit Great Gull Island, a major conservation and research center for terns out at the mouth of Long Island Sound. The island was only half a mile long, seventeen acres in extent. It lacked trees, except for a few apples and red cedars. The original growth would have been held down by wind and salt spray and was suppressed later on by the army. A massive fort was erected there to protect the entrances to New York harbor, beginning with the Spanish–American war, from enemies who never put in an appearance.

The common terns that now occupied it nested on surfaces of peeling concrete that covered old bunkers, as well as on stony ground through patches of weed and grasses. Another species, the roseate tern, made its nest between huge piles of trap rock along the shore. Night and day, they cried and chased each other during the early period of courtship flights and pairing up. It was an out in the open, exposed, and classic exercise, while the abandoned fort, with its concrete walled ravines banked with honeysuckle and its dark, underground tunnels, waited for more centuries to wear it down.

There were millions of people not too many miles away. Jet planes often droned overhead, and fiberglass boats cut the choppy waters beyond the island shore, while the great oil tankers moved blindly by. But there were no newspapers, no radio, no TV, no electric light, no hot and cold running water. It was, strictly and simply, tern country. The place was allowed to belong to them. There was "nothing to see" out on that relatively isolated island but the urgencies of life stripped to essentials.

The Great Gull Island program was under the direction of Helen Hays, of the Department of Ornithology at the American Museum of Natural History, which owns the island. She is a great enthusiast, with a strong vein of practical determination, and manages to keep any number of aides, and even temporary visitors, on the alert and busily working from dawn to dusk. She and her co-workers have spent many years collecting and analyzing information about the twelve thousand birds that nest there. I remember, after trying to sort out what seemed at first like wildly unconnected behavior patterns in the courting birds, that Helen said: "Your questions are beginning to improve." That encouraged me. I felt that I was starting to circulate, perhaps moving in on the ways of migration itself. I had never quite understood why the migrants arrived each year at about the same date, but dates are a tyranny enforced on us by human history. Long distance migration seems governed by internal rhythms in the birds themselves, which are coordinated with the motions of the earth.

I arrived at Great Gull Island on the second of May and was told that a few terns had come in the day before, as the Russians were parading in Red Square, but had left at sundown. As another visitor and I were landing, making four people on the island, we saw a number of the birds flying out across the water. They also made brief overland flights and then moved off again. With their sharply cut wings and dancy, limber way of flying, some were paired up in courtship flights in which they might flutter up together, or swing like pendulums, alternating, one above the other; but most were dividing up into temporary trial flights, in which three or four would be engaged, but soon broke off. Twittering barn swallows, which had arrived earlier, were also chasing each other over the buildings of the fort.

At seven p.m. a long line of about sixty terns passed low over the water parallel to the island shore and then doubled back. They engaged in this action seven or eight times. It reminded me of the sideward sweep of alewives as they came into a shallow stream out of deeper water, or of dovekies trying to avoid land but getting swept in closer and closer during a winter storm. Like other seabirds, terns have an innate fear of the land and its predators, which they have not yet overcome at this stage. Normally, they do not land and stay overnight until a week or ten days after these preliminary maneuvers. Each day they leave with the sinking of the sun into the sea and do not return until dawn. Their hesitancy looks like that of a scouting party not ready to tell the main group to come in, although it is probably as much a matter of the accumulation of feeling in them, the gradual pressure of numbers and example. Eventually, enough would settle in and start making the motions of nesting, acting as "anchor men" at the various sites on the island which seemed suitable to them. For the time being, it was a matter of approach and withdrawal. So as the surrounding waters turned pinky and opaque, having received the sun's last bloom over their purple meadows, the terns sped away over Long Island Sound to roost along other shores.

They had shown up a few days earlier than expected this year. It was likely that the food supply, the small fish they fed on, was abundant and that this was one of the reasons they began to move in when they did. They may leave the shores of South America in late March or early April, although not a great deal seems to be known about the time it takes them to make their northward migration. They beat ahead, in small or medium-sized flocks, bending to the marine weather, occasionally diving for what small fishes are available along the way. The relative consistency of their annual arrival off New England from place to place is hard to grasp in detail, as if the migrants followed the seasons beyond our knowledge of them.

There was always the unlikely possibility for those who waited for the terns to arrive that they might not settle in. Catherine Pessino, of the American Museum, said that the early days before the island became a full-fledged research station, back in 1965, involved periodic visits to keep a record of the number of terns as they started to come in. They did not arrive in good number for some years, and the evenings spent in waiting for them to show up were nothing short of prayerful. Even now, after many successful years, it was a somewhat anxious business. If the birds failed to move in, the loss for those who served them for years on end would be like the loss of light. Back to the dungeons!

On May 3, at 6:45 a.m., I counted forty-four terns on the dock area. They were roosting on timbers erected for pilings of an old wharf no longer in use, as well as on a smaller, adjoining dock, where the boat from the mainland arrived and departed once or twice a week, depending on the number of visitors. At least one pair was posturing in courtship, the male holding a little shiny fish in its bill, both thrusting their heads and necks up at a sharp angle, tail feathers cocked, and wings held out a little to the side.

By 7:05 a.m. the numbers in the dock area had built up to seventy-five, and in two more hours to eighty-seven. This did not count others flying in over other parts of the island, but it amounted to a majority. It was hoped that a thousand or more might appear in a few days.

A thick, early morning fog cleared off by noon, but it had turned into a generally cloudy day. Instead of tentatively flying in and then leaving, the terns were now staying over the island for longer periods. They clipped easily through the air. They circled, fled out over the water, and returned. They cried harshly, persistently. The "keeaarh" or "keeurh" calls by which the commons are known can also be heard as a nasal "ayhhurr-ahn-ahn-ahn." You hear "chivee" and "chuwick" from the roseates, but like all our renditions of what birds sound like, these few calls do not take us very far. Tern talk is self-relating and exclamatory, a condensed, punctuated expression of the emotions. Most of its silvery, strident tones are obviously lost to the human ear, but it is possible to put some intentions into them. They land, fly off, threaten, entreat, and their cries change accordingly. They are also so much a part of the changes in their surroundings that I am sure the tone of their calls must shift in subtle ways in compliance with them.

As they flew back and forth between land and water, I was conscious of their restlessness, and of how quickly responsive they were to one another. At 12:40 p.m. the number on the dock had increased to one hundred sixty-four. A northwest wind had been gathering strength during the morning, and increased to gale force later in the day. By mid-afternoon the birds had all left the island. The storm might have discouraged them, but at this early stage in the nesting season they normally leave by evening in any case. And they are all-weather birds. I have seen them out at the tail end of a hurricane, beating ahead, rising, falling, twisting like leaves, and strong winds are not stronger than their drive to move in and nest. One spring, however, there was a wild, wet snowstorm on the ninth of May, and although some terns had already started to settle in and make nest scrapes on the ground, they seemed to have felt the storm two days before it hit and cleared out. They are very sensitive to atmospheric pressure, quickly alert to changes in light and temperature, as, for example, when the clouds move over the sun, or the sun sails out again. This was an island of freakish weather as compared with the mainland, and the terns, being nervy, excitable birds, seemed to respond accordingly. Fog moved in; the sun burned through; the wind shifted; rock-gray, cloudy bands started running the horizon; then raindrops fell out of a thick overcast like gray wool; and the birds kept responding, through their ever moving natures. I came closer to the weather myself, thinking of synchronizations so fine we can't readily detect them, feelings so passing quick we are unaware of them. It is the way of earth. An animal nature alert to no more than a flying cloud has much to teach us.

During the morning I saw two Canada geese flying toward the island. When they reached it, they suddenly saw it as an unfavorable place to land and so veered off and banked away. As they disappeared over the water, I felt the searching in them. What a miraculous planet, to bring out the qualities of discovery in all its people!

Very few terns had come back this day, although the wind abated to some extent, after blowing hard for three hours. At 6:30 p.m. I could only count four. An hour later, twelve flew in from the sound, at about one hundred feet, but failed to land, speeding off over the water again.

After dark, we saw bioluminescent organisms shining in the

seawater as it plunged against the rocky shores, each one as conspicuous as a star in the heavens.

From my sleeping bag, at waking intervals, I could hear the wind blowing all night long.

On the fourth of May it was cold during the early morning with the wind blowing from the north northwest. One flock of a hundred terns and another of ten flew in but left again. More showed up later, coming in out of a watery distance tossed by the wind and blazing with light. There were eighty-five on the dock area and another forty in the air at 9:20 a.m., but they only stayed on the island for about an hour.

Away from the wind, the barn swallows sliced at flying insects across an enormous, round, empty gun emplacement, one hundred feet across, which had once housed a "Big Bertha" of a coast artillery gun. It was Roman in its proportions, the kind of dominating structure human society attains out of its organization and thought. "Cultural evolution" builds greater and greater nests, monuments and platforms from which to rise and fall, while the proportions of the swallow's nest remains the same. It is an artifact that outlasts the world.

Based on past years, Helen told me, it could be estimated that the terns would not spend the night on Great Gull until the tenth, getting down to the business of nesting by the fourteenth, but this year might be different. We spent hours weeding along the shore so as to make more nesting space available, especially for the commons, since the roseates like thicker vegetation, or nested deep between the crevices of the trap rock. Eating supper by candlelight, we made bets. How many would there be tomorrow morning? On what day would they start to move in for good? Nothing seemed more important.

By six in the morning on the following day, May 5, about two hundred arrived, and in another hour, six hundred fifty.

During the past few days, other migrants had come in too, most of them, with the exception of redwings, song sparrows, barn swallows, spotted sandpipers, and a few other residents, stopping over briefly. Various finches and sparrows, catbirds, thrushes, and warblers will stay for a few hours, or a few days, depending on the weather, or the available food. Three sharp-shinned hawks lingered for two days on their way north. Two of them left, having been a cause of great unhappiness to the terns. The third was followed over the island and mobbed by them in the morning. It dropped down into the thickets, to reappear several times during the day before flying after the others.

So many terns had now come in that it seemed clear they meant to stay and start nesting very soon. Delays caused by stormy weather only increase their impatience. I could hear and see it in their actions. Their courtship parades on the ground had intensified. Their mating flights were more urgent and sustained, and they passionately cried out all morning long. They led and chased each other, and their elbowing, gliding, quickly shifting wings hardly ever stopped. Their energy seemed incredible to me. It is said that they obey a sexual impulse that is confined to a short period of the year, a biological change that may be set by going by the action of the light and is reflected in the glands and gonads. This makes good biological sense, but where all that zest and hunger, that restless dancing comes from escapes our reasons for it.

To a casual onlooker, their game might look driven and nearly self-exhausting, a life made to go for the limit through no fault of its own. It is a grace in action that is to be admired but not envied. At the same time, these birds are as hurried, tense, angry, and petulant as the rest of us. Watching them during other busy mating seasons, I also got the impression that they could even be quite aimless and forgetful. This could be the result of a limited intellect, but since I experienced the same traits I could only feel that we were all in this together.

The wind, on the morning of the fifth, had come from the

south, bringing warmer air, but the seas were moderate. I heard a roseate overhead with a shrill "keep-keep" of a cry, which might have been a reaction to the presence of the sharp-shinned hawk. A flock of terns, white and gray as the clouds, lifted up from the shore in sudden silence, flew out over the water, and then returned, crying out again as they landed. These have been called "alarm" or "dread" flights, or simply "outflights," but it is an exercise in feeling, a kind of communal, nervous spontaneity, and might also reflect the continual conflict in them between land and water.

The wind had picked up during the afternoon, and in a couple of hours it was blowing at gale force. The terns roosting along the shore were all hunkered down, facing in to it. By evening they all left the island as before, flying out over heavy seas. Because so many more had come in during the morning, much was expected of the following day.

On May 6, the final day I was to spend on Great Gull Island, there were six hundred terns flying over at 6:00 a.m. The number roosting along the dock reached two hundred fifty by 9:30, and in another few hours four hundred. So at least a thousand birds had now come in and were either roosting along the shore or racing back and forth in their air. Occasionally, a pair would go off together in a wing-dipping, floating flight, measured and assured.

I spent some time watching the roseates, not so numerous as the other species, less strident and agressive, and a little later in arriving. Their breasts, shining clearly through the misty, morning air, had the faint, suffused pink during the breeding season that gave them their name. It is a color of sky and water, sunset reflections in the clouds. Their wing feathers are a paler gray. They have sharp, black beaks instead of red ones tipped with black like the commons. Their jet black, glossy caps stand out handsomely against white breasts with that subtle shade of pink which is not always visible when in the wrong light. Their white tail feathers are deeply forked, with the two extended "streamers"—longer than in the other species—pronged out behind them.

Some of these fine birds were posturing together where they were lined up with the common terns along the dock. The fact that they did not have much room for maneuver only gave their performance an added style. They managed it like a matador with his cape. They would weave around each other in tight little circles, bowing and craning heads and necks, holding their tail feathers straight up behind their backs like spars, with their wings bowed out at their sides, and the tips of their primary feathers crossed over. They kept up this stylized performance for a few seconds and then stopped, having established a relationship with each other, and a set distance. Roseates are more rigid in their courting and mating than the commons are. Their ritualizing is more set, implying a fierce, almost steely order, behind it. Watching them, I thought of ice, the cold, classic beauty of it as it begins to visit brooks and streams in December, taking the form of diamonds, ferns, or, in that strict, surpassing way that forms conspire over the earth, feathers.

Toward sundown, as the sky grew soft with mackerel clouds, hundreds gathered along the shore instead of flying out over Long Island Sound. We watched a few pair in open nesting areas, some distance from the others, as they made pre-nest scrapes in the gravelly ground. They did this with breast forward and down, bodies tilted up in the rear, circling a little while they constructed a shallow depression with their small, webbed feet. I could see one pair already going through the motions of copulation, the male on the female's back. These few birds seemed tentative in their movements as the day waned, as much drawn to rejoin the greater number roosting around the periphery of the island, where they conversed and preened, as to the nesting places they were now trying out. But they had the pioneer influence, and as darkness covered the island we felt sure that the terns now meant to stay.

Besides their outflights, which took the form of a noisy panic when a predator like a hawk came into view, or were occasionally performed in silence, the terns would also swoop out over the water at dawn, or they would fly up from their nesting areas when a passing cloud shut out the sun. They were very responsive to such changes. One evening, about a half an hour before sundown, at that special time when daylight begins to lose out, I saw many fly up from their roosts and head out across the sun's golden path, where it shimmered across the coldly blue, choppy waters, and then swoop back in to land again. The sun had a flaring, flowery corona of gold-orange over little drifts and thin, dark rafts of clouds lying above the horizon; and when it changed to more than a streak of blood red sinking into the sea, the birds flew up again, to hover and mill around like burning chips in the sky, as if to offer an evening service in homage to the light. And so I realized the splendor of the sun through another vision than my own.

The writer Michael Harwood, on a visit to Great Gull Island, asked a young student assistant who was on his way back to the mainland how it felt to be going back to the real world. "But this," he answered, "is the real world." It is not often that we can shed the unending distractions with which we surround ourselves, but when we do the realities appear.

CHAPTER 8

The Ring

*O*ur society seldom respects what it does not know, or cannot claim for itself. We record and analyze things down to the unseen atom and the genes, while interposing a vast world of manufactured fantasy, married to the destruction of resources, between us and the life around us. The life-sharing mysteries of love and balance come in unannounced, as if from another world we had thrust aside.

The magic arrives through snow and rain, lightning, sunshine, and shadow; it sways, flickers, and shifts with the light and suddenly announces itself through a gust of wind in the treetops, or in the call of a bird. When the haunting cry of a loon comes over the water, a great company, which had almost moved out of sight and hearing, seems to fill the morning. We need the loon not only because it is a rare bird but as a link with primal reality. Rarity implies dead ends, and if there are too many of those, none of us is going to be able to see very far into the smoke we are creating. The loon's wild eloquence and clarity fills me with awe of the world we never made.

A man in Newfoundland by the name of David Quinton once told me a story that had to do with the disappearance of an old woman while she and her family were picking berries by the shores of a lake. Later on, he was told by a young man and his father that they had met an old woman with a sack of berries on her back in the same vicinity. She had suddenly appeared "out of nowhere" and warned them off. He himself was walking there when he was petrified by the sound of a woman yelling, but shortly afterward a loon came into view, calmly riding the waters in its stately fashion.

The spectacular voice of the loon, beautiful music to those who love the northern reaches of the world, breaks into the human ear with news of a vast interior. As Richard K. Nelson relates it in his book, *Make Prayers to the Raven*, about the Koyukon people of Alaska:

> Stories of the Distant Time say that Koyukon songs originated partly from the loon's crying and partly from human imagination. Perhaps this too is why loon calls are especially meaningful. After the late elder, Chief Henry, passed away, I was sitting near his house with an old man from Allaket. "It would be good to hear *dodzina* (the loon) right now," he said quietly. "We really like to hear that music." He had tears in his eyes as he talked about that beautiful wild voice, about hunters listening to it as they paddled their canoes from lake to lake in the spring. And he seemed to wish for it now as a good sign or a gift from the natural world to the spirit of a great hunter. I could scarcely have troubled him to explain further.
>
> A few days earlier, while Chief Henry lay near death, an old woman from a Yukon village walked to the nearby shore of Binkookk's Lake. She stood at the water's edge and sang Koyukon "spring songs" to a pair of loons that had been in the lake for several weeks. Shortly, the loons swam toward her until they

rested in the water from some fifty yards away, and there they answered her, filling the air with eerie and wonderful voices. When I spoke with her later, she said that loons will often answer spring songs that way. For several days people talked of how beautiful the exchange of songs had been that morning.*

The cry of the loon belongs to those proclamatory circles that widen outward on the universe, like rings on water spreading from a drop of rain. The other name for the common loon is the great northern diver. So the human soul is the great diver below the waters of experience.

Loons are suffering from the same pressures that have restricted or eliminated so many other wild inhabitants. Regions that were their ancestral home have been invaded and often violently disturbed. The solitude they love is nearly gone. Their nesting places by the shores of once quiet waters have been interfered with by housing, summer populations, motorboats, and in some instances the damming of outlets, so that water levels fluctuate and make their water-edge nests untenable. The loon's breeding success has dropped accordingly. Those wintering in coastal areas have often died as a result of oiling, and no one knows the long-range effects on them of the contamination of the fish they eat.

The loon originally bred in most large ponds and in lakes all the way across the northern tier of states, from Maine to Alaska, Oregon, and Northern California, before so many were shot out and then became victims of progress. It is a large and radiantly handsome bird. With its rich, velvety green and black feathers marked with white, it reflects the dark waters and the evergreen banks above them.

One July day, while we were out canoeing on a small Maine lake, Frank Burroughs, Jr., and I met a pair of common loons, with a tiny chick between them trying to climb up on its moth-

*Richard K. Nelson, *Make Prayers to the Raven* (Chicago: University of Chicago Press, 1983).

er's back. The male sent out a long, quavering, ringing call as we came closer. He kept his body between us and his family, making a great demonstration of spreading his wings and rising off the water on his tail, showing his white breast. Then he dove, and while we were wondering where he would reappear he suddenly bobbed up out of the water only a few yards from the canoe, with an ear-splitting cry, and then retreated. I have never been so conscious of that wild ring of protectiveness with which animals surround themselves; it seems to circle back to a time when we were scarce.

We force the worlds of life into a corner. They become a sacrifice to our monopoly of space; but all true sharing, and interchange, lies not in us but in the unseen allowances of wilderness space. Listen to the music and you hear the earth's designs.

The outer beach of Cape Cod is often covered by more people during the summer than it is possible to count. It is also occupied by a few, small, scattered colonies of least terns, once abundant, now in the rare or threatened category over a major part of their coastal ranges, east and west. They are lovely, gentle little birds, with a tinkling quality to their voice. They have yellow bills, black caps, soft, ash-gray feathers that blend in with sand so that they are hard to spot; and they have white foreheads that must be helpful identification marks for them, since they shine like signals in the light.

Like other species of terns, the leasts are nervy and volatile, quick to react to intrusion; too much of it, and they may desert their nests. Human beings, stray dogs, and predators may keep them in the air too long, so that eggs or downy young will cook or die in the hot summer sun. Dogs can cause terrible trouble in a tern colony, but their owners are sometimes as difficult to restrain as they are. They can be fined for not having their pets on a leash, although this is not a heavy burden for most people who can afford it, or even when the dogs chase deer, but there is no penalty when they disrupt or destroy a tern colony, which gives you some idea of our priorities.

When they are only a few days old, the chicks will run out from the nest, only a little scrape in the sand, at the approach of an intruder and head for the nearest hollow. If this happens to be a rut made by beach buggy tires, they are in great danger of being run over. So in those areas that are enlightened enough to hire wardens to protect them, their little colonies have to be roped off and signs erected to warn people and vehicles away. Tern wardens have to exercise especial vigilance and try to educate everyone they meet as to the vital need of protecting these waning populations. At times, this almost looks like an attempt to hold back the tide. When the need is patiently explained to people, they usually sympathize with it, but hundreds of thousands of others will never know it exists.

One morning I accompanied the warden on his rounds while he checked on his signs and counted his charges. The bathers, not far away from one of the colonies of perhaps thirty birds, were spread out in groups, couples and singles over the sand. Coming on them through the haze of heat, they looked like some strange breed of sea mammals, but with an abstracted gaze as they looked out over the horizon like nothing else in nature. What could those nearly invisible birds, nesting in the sands, possibly make of his major intrusion on the world's habitats!

As we progressed to the next nesting site, we saw a horse van, with two gray horses waiting just beyond the Do Not Disturb signs. Not long afterward, two beach buggies came lumbering awkwardly down the sands, trying not to get stuck in the soft places. One carried a camera crew, the other a load of young women. When they stopped by the van, one of the girls jumped out first, with a brilliant Pepsodent smile, and started to pose next to the horses. What all this amounted to was fifteen minutes of highly paid shooting to advertise a special brand of blue jeans.

Only a few yards away, the unnoticed terns hovered nervously over their chicks. They were more aware of the humans who had encroached on their territory, that vital stretch of sand, than the humans were of them. What odd beings they were, roaring by in their machines, or overhead, milling aimlessly around, occasionally blundering in, idly curious, and then retreating, until the darkness brought the beach back to the sound of the waves, and an occasional fox stole down from the dunes to see whether it could find some eggs to eat.

Within the earth's great circles of atmospheric movement, of sea change and sea music, with long distances in their minds, each pair guards its round bare nest. After the young are hatched, the adults spend weeks intently feeding and protecting them. (When the weather becomes abnormally hot, and the sands blistering, so that exposed chicks might easily expire, the common terns, larger relatives of the leasts, have been known to wet themselves in the salt water and then fly back to cool them off. I am not greatly moved by what we as a race do to protect ourselves, although it arouses me at times to extreme apprehension, but here, in these nonhuman circles, I feel a common home.)

During a spring visit to the island of Nantucket, I followed Nan Jenks-Jay on her rounds as a term warden for the Trustees of Reservations. A downy, buff-colored least tern chick had wandered away from the nesting area guarded by its parents. It went tripping off toward the line of wrack and driftwood on the upper beach, where she picked it up and carried it back.

"You've got a lot on *your* mind," she said in loving tones to this wayward child. What wonders, I thought, are located in that downy body. Who could know what is on its mind? It belongs to a world of profound perception and impulse, of a variety of detail we only come in on from the outside, of small particles, rays of light, subtle changes in the air, unseen signals, a continental recognition in the mind. Might not this innocent with its wide connections have as much right to intelligence as I do, certainly a right to the circles of love? The universe is not concentrated on the success or failure of the human experiment but on the birth of a bird.

CHAPTER 9

The Travelers

To impose our one-sided requirements on earth's living space and then expect to be accommodated presupposes a knowledge we do not have, and an ability to change course that we are by no means sure of. Not even the advanced numbers game we are playing can tell us just how much space any given life and habitat needs to be sustained—far more than we allow them, it is fair to say. In fact, the more we become dependent on the use of numbers the less we seem to know. There are environmental firms that can be called on for consultation about land for conservation, watershed areas, and other purposes. They draw up reports on the basis of computer-fed, statistical analyses of ecosystems; but there have been instances when the people whose job it was to interpret these masterpieces to their clients were unable to tell one form of life from another. What is this "environment" we talk so much about? Who knows it?

Even those who study the variety of species and their specific relationship to what the earth affords them do not necessarily know what they need. After all, the arctic tern, traveling up to eighteen or twenty thousand miles a year, has to have its circuit of the planet. Other members of the tern family are birds whose lives are rhythmically involved with oceanic currents and atmospheric conditions whose nature shifts in emphasis not only according to the season but from week to week, from day to day. Their food supply is not consistent over any given area, but has to be hunted over a wide range. The fish for which they hover and dive in surface waters depend for their existence on an exchange between the oceans and the nutrient-rich salt marshes, inlets, and estuaries along the coastline, and all the streams and rivers that run down out of the land. Each species of tern has its particular habits. A roseate may perfer fishing out on shallow waters over sandy bottoms. A more opportunistic common will work closer inshore. None of them, however, can be restricted to a narrow range. They have to have an open, viable, and fluid space where they can devote their lives to searching for its advantages and accommodating to its risks. Where the terns lose out is when they have their flexibility taken away from them, not because they are too narrowly adaptable. This is a prospect the human race may be facing.

Still, that great complex, roving distance with its multitude of orbiting lives is our only source, our home, and we depend, as Buckminster Fuller put it, on its "competence." Out of it, season after season, come the true navigators, the global runners, those who sense the crossways of the earth within themselves.

The seas well up, with their dragging tides. The clean waters splash against the rocky prows of outer islands along the Maine coast. Female eiders herd their groups of young over the water, where a loon rides like a carved boat. These islands, with their ledges of tortured and buckled schist and granite, are occupied in spring and early summer by various birds that move in and make their nests, such as the osprey, great blue heron, cormorant, laughing gull, and common tern. Perky song sparrows call from the rock and the clumps of seaside golden rod, or bunches of sea lavender along the shoreline. Rafts of seaweed come floating by. Schools of fish run and circle in the surrounding waters, searching them for food. Gulls cry out in hoarse and defiant tones. Towering clouds gather and separate thousands of feet up, while the islands are rocked into the music of water, wind, and birds. Life is balanced in with the weight of the seas. It belongs to a sky thrown open, while the waters, under the determinations of air and light, keep roaming and sweeping ahead. There is no letup to these great motions, and they pull a response out of every living thing. The islands are magnets, crucial centers for the will to subsist. They are testing places for those travelers who move in each year to breed and nest and bring up their young, with everlasting commitment. Others stay in place, like the salt-tolerant plants, the stunted spruce, or they move on to other hemispheres and come back again.

What is it to the arctic tern, now only occasionally seen in these latitudes, that we have reached the moon? It is still exploring the earth's rhythmic distances, and in that sense it is ahead of us, out on the open world. In the far northern lands where it breeds, the arctic knows only a few hours of darkness at the most, and during the Antarctic winter in feeds between the ice floes where the sun fails to set. So it is a bird of light, or one that flies around the globe ahead of the dark.

"Who am I?" I ask the migrants, and they never answer, unless by moving beyond my horizon into the currents of the seas and the atmosphere they signal the depths of the question, and so begin to move me ahead, while I am still standing on the banks of a stream, looking in at the moving water.

One year a group of migrants out of coastal seas but rarely seen close to shore appeared off a Cape Cod beach after an April storm, which had apparently forced them in. They were phalaropes, in gray and white winter plumage, scattered in the shallow waters on the rim of the sands. They were twisting back and forth, half swimming, half walking, while being lightly flung back by lines of wavelets agitated into white caps by the wind. It was a kind of joy ride on minor turbulence. They would flit up from the water and almost simultaneously land on it again. They turned and dipped, dipped and turned, with a nervous, graceful agility. They were constantly bobbing their tails up and down, to keep their equilibrium, while pecking at the water with their beaks for food.

How was such a bird able to manipulate a body that seemed at cross-purposes with itself, a body that ought to be shaking apart at both ends? But what looked like a constant jerky turning would switch, now and then, into the perfect spinning motion of a top. They would also interrupt the busy beauty of their actions to make aggressive, crouching little rushes at each other, squeaking and peeping at the same time. They were like animated toys, with their bustle and hurry, bob and spin, turning at intervals like a feather on a fulcrum, magically adept; but it would be a mistake to see too much fragility in them. The red phalarope is a pelagic bird, and the related northern phalarope somewhat less so, but both are often seen some miles offshore, at times in the company of whales, which help them locate the plankton they feed on. When they flew off they looked strong enough, like many other water and all-weather birds, in what they had to do. Flying off seaward into the wind, they beat rapidly forward with a kind of bold ducking and dropping technique that showed they were well equipped to meet the turbulent waters ahead.

Such compositions of delicacy and strength bring land and sea together. On a spring trip to the shores of Hudson Bay, a thousand miles to the north, I saw these birds again. Cold winds flew over a vast, reaching landscape of wet muskeg and tundra, with dwarfed tamarack and thin, spindle-shaped black spruce. When we set out in the morning there were frost banners on the trees. The cold air poured out freely and settled over the open spaces, while pockets of vegetation held the warmth of the sun. A thin layer of peat and root-bound soil covered a pavement of permafrost a foot below the surface, and it was covered with cushiony mosses and lichens, in golds and browns and greeny grays. Rosebay and labrador tea were coming in to flower. Their growth was synchronized with the speed of spring, freeing the land from its winter vice, and they seemed to have a readiness of their own, taking advantage of the light while there was time.

The nesting birds, with their own intimate association with continental changes, were starting to nest in islands through-out the muskeg, on hummocks, or in grassy areas free from water. Their eggs had the wonderful colors of the wild, blending subtly and secretively with the ground. The eggs of the semi-palmated plovers were reddish and speckled; those of the stilt sandpiper were greenish black, with marbled patterns; the whimbrel had gull-sized green eggs; and the least sandpipers were a light, leafy red color. A wild bird's egg is a great art, as perfected as a water-worn green stone by the seashore, or a veined leaf. There is nothing like it for belonging, and at the same time the purpose it serves requires a uniquely delicate disquise. It hides, under the earth's protection, while it is put in supreme risk in a predatory world. It is a design of infinite daring.

Watching those nesting grounds through my field glasses, I saw a phalarope very quietly leaving its nest in the grass. There I found four olive brown eggs with darker brown markings, their pointed ends all facing inward. After I had walked back and waited for a while, the bird returned to the nest, walking so silently and unobtrusively that it would have been invisible if I had not seen it and known where it was headed. Later on, I watched a pair of phalaropes out in the open as they circled in a shallow pool, spinning around as they pecked at the water with their bills for the organisms they were feeding on. So the mating dances they had introduced me to were joining with the vast, cold upper reaches of the subarctic, with its sea fogs, its blinding snowstorms, even in June, and its relentless, elemental balance.

The fabulous nature of long-distance travel in animals seems to be outside our common experience. We take it as a stunt. Modern life is surrounded by all the aids to navigation it could possibly need, and so we forget how to navigate. To find the reasons for the ability of a tern to travel around the world and return to its nesting grounds, we look for some specific mechanism to account for it. Homing becomes not so much a common, if mysterious, property of life within the body of the earth, lying buried in ourselves, but something to be accounted for through science. People who have been displaced from familiar locations and then blindfolded have been able to point their way toward home as well as any pigeon. It seems that there is a whole world of directions in us going unexercised. The fact that the long-distance migrants in nature often seem beyond us, and more spectacular than anything in the circus, might indicate that we were becoming strangers to discovery and in need of rescue. How shall we find our way to the future unless we, as well as the birds, can locate ourselves by means of the patient landmarks of the earth, as well as the sun and stars, the wind and the rain?

Another astonishing migrant is the wheatear, a perky, sparrow-sized bird of stony ground that nests in the tundra, laying pale, bluish-green eggs in a fur or feather-lined nest of grass in some crevice or abandoned burrow. During the postglacial period it nested in Britain, having so to speak, been held back by ice for thousands of years. Then it gradually extended its breeding range out into the western Atlantic over stormy seas to Greenland and northeastern Canada, and it is still moving farther to the west. The wintering grounds it returns to are in Africa. On the other side of the continent, wheatears reached Siberia and then crossed over into Alaska. They are now extending their range across eastern Canada, so that they have nearly met fellow migrants colonizing it from the other direction. This is not just a little bird on "automatic pilot," but one that has followed the seasons of earth history. The scope of its adventures takes one's breath away.

The race of wheatears nesting in Alaska make their return migration flitting along at about eighteen to thirty miles an hour, over the vast stretches of northern Asia, and southward across the Mediterranean to winter in semi-desert country south of the Sahara. Their scientific name, oenanthe, comes from Aristotle and means "wine flower," since they arrive in Greece when the vineyards are blooming. The progressively longer migrations this bird made after the last ice age, so that it travels for eight months of the year, like the arctic tern, proves that it has only the limits of the planet to confine it. This phenomenon may seem pitiless in terms of those universal energies that require it. On the other hand, it is as close in its need and in the way it satisfies the distinctions of space as any plant or insect in our backyards. The wheatear, like all other migrants, near or far, keeps searching for its own places, its distinctive lands of bare, sandy, or stony ground of unimpeded light, which have the seeds and insects it feeds on. This little bird carries the recognition that the whole earth insists on for its survivors.

Migration for these animals is not simply movement from place to place but a life and death journey in tune with the motions of the planet. Their whole physiology is involved. In his study of Hudsonian godwits,[*] Joseph A. Hagar found that

*Joseph A. Hagar, "Nesting of the Hudsonian Godwit at Churchill, Manitoba," Ithaca, N.Y.: Cornell Laboratory of Ornithology, 1966.

this large, long-legged, highly specialized species arrived on its nesting grounds in the Canadian tundra in late May or early June. A first clutch of eggs might be laid between the twelfth and fourteenth of the month and the chicks hatched out in about twenty-two or twenty-three days. Adults nesting in the Hudson Bay area withdraw and move down the coast in late July before they head for their wintering grounds in late August. The young leave about a month later. Then they disappear from sight, flying nonstop at an estimated height of twenty-one thousand feet above sea level, until they reach Argentina and Chile.

Accelerated growth in the young is synchronized with the dramatic speed of spring in the subarctic, before the long days of a short summer and its abundance of insect food.

> The central fact of the growing period in the Hudsonian godwit and, indeed, of its whole nesting cycle, is the extreme compression of every corner to save time. Arctic shorebirds as a group have developed a pattern of fast growth beyond most birds; the godwit completes its cycle among the first, and all things considered, is probably outstripped by none.

These migrants are outreachers, not passive dependents. They measure an exalted risk, as the spring light, quicksilver on the trees, gold on the waters, challenges every living thing, as violent as winter in its demands of life. It is not so much that a Hudsonian godwit, or a monarch butterfly, has no alternative for what it does, but that it takes part.

The kind of unplanned and nearly uncontrollable conditions man is letting loose on the planet is leading to the end of innumerable species whose lives have always been staked on a finely balanced relationship with the energies that both threaten and fulfill them. They are highly vulnerable to an outside force that stakes everything on the assumption that it can bypass these abiding relationships, that does not have to depend on precarious food supplies, or nest at the thin edge of winter, or wait on the tides, but is able, temporarily at least, to change entire environments to suit its insatiable needs. That we too are vulnerable goes without saying. The thought is terribly with us, night and day, the more so when we only have ourselves, imprisoned by our own assumptions, to show the way. But there are others, if we would be followers, to lead us toward the kind of sharing the earth has always demanded, according to its uncompromising and beautiful standards. Birds, fish, mammals, plants, natural habitats are fast disappearing not only because of the murderous stress of the human presence, but because they lack enough company in the human spirit.

CHAPTER 10

The White Pelican

When I climb into my car and head either northeast for Maine or south toward Florida, I can avoid the dangers of biological flight between the Caribbean and New England. I can reverse my migratory heading any time I want to. This is the way freedom from elementary peril and need, freedom from nature, is supposed to have been achieved. But although I enjoy the relative freedom of owning a car and going where I please, I recognize that it was made possible by a civilization that is highly concentrated on itself. It may be, after all, that some of the anarchic and specialized ways of our society are biologically preordained. The ants go their own way, too. Part of the reason for their success, as William Morton Wheeler pointed out in his famous study, *Ants*, is due to the comparatively small number of their enemies: "As Forel says: 'The ants' most dangerous enemies are other ants, just as man's most dangerous enemies are other men.' "* So we do not really avoid danger; we just invite our own accidents.

Like the ants, I have to stick to the roads we have constructed, a continental web of them. It is no easy matter to stop on some of the superhighways without high risk; but if I were unable to get out and walk somewhere along the way I might never see a thing. When driving, I am not aware, as I might be when trying to cross a highway on foot, that my car goes tearing by like a maniacal bee deprived of its senses. For the most part, I just tamely stick to the route and get to the other end of it, as part of a regulated environment, insulated from what it goes through. Travel does not mean going out over stormy seas to reach unseen shores, or fishing for our lives, but joining a network. So when I move out now and then, leaving the barnacles behind me, I take our messages with me, transmitted by the car radio. "Join us in this great crusade," says a built-in evangelist at the end of his harangue. "See ya."

Birds travel the earth as a different breed of searchers, at times on very direct routes between their wintering and breeding grounds, at times circling to find their way over immense distances. They have the planetary currents, the major tides of weather on their wings. They are often blown off course. The great space that is intrinsic to their lives is also the space for getting lost. At times, I wonder if that is not the kind I am looking for, not so much to get away but to enter in.

In 1974 a white pelican, which may have soared in from its breeding grounds west of the Great Lakes, having been forced eastward by a major autumnal weather front, spent much of the winter in the Herring River and reservoir area of the town of Harwich on Cape Cod. I first saw it just after Christmas, and it had frequent visitors for as long as it stayed. I last saw it well on in February, before the water was completely frozen

*William Morton Wheeler, *Ants* (New York: Columbia University Press), 1910.

over and it had to go somewhere else for its food. For a while, some fishermen who chopped holes in the ice, where they dropped in their hooks and lines, slung it a few fish, and the bird readily accepted them. It used to stand on a small, grassy hummock above the ice and preen. I watched it flapping and gliding on its nine-foot wingspread very high in the air until it became almost lost to sight. Clearly, it was a major, continental bird. When it flew up off the water, still open around the edges of the reservoir, with a few herring gulls screaming at it in exasperation, its white excrement streamed down. It had a long, pinkish yellow bill and pouch, yellow-orange legs and feet, and the solemn look of a tragedian indifferent to tragedy. Trying to get a picture of it with my camera, I skirted one end of the reservoir, stalking the bird through the woods and thickets along the shore like a phony Indian; but it spotted me long before I could get close enough and casually flew off to the other side.

I set off for the south on my own voyage at the end of the winter, feeling a little like an "accidental vagrant" myself, when I got temporarily lost on the fringes of the great city I had once lived in. I wandered in and out of a district of empty warehouses, driving under half-built overpasses and down side streets that shouted at me about all the tense concerns I might have forgotten by moving to the country. What is nature but human nature? Then I joined a jostling, semi-congealed mass of computer traffic trying to work its way beyond the city during the rush hour. Under these conditions, the automobile loses the freedom of the open road and at the same time is one's security against chaos. Thousands of us, crammed together, only drive and wait, in a kind of enforced patience. Perhaps fury and weariness just cancel each other out. All the complex structure and greed of a civilization is concentrated on an individual sweating out a traffic jam.

I can get inside a newspaper, or restrict my travel to an

expressway, the manmade equivalent for the pelican way, or what the Anglo-Saxon poem "The Seafarer" called the "whale-way over the stretch of the seas," and experience the equivalent of shutting myself up in a room. In traveling so as to make time, or cheat it as much as possible, I become part of the spreading, gray-brown discipline of the highway culture, spending each night at a stereotyped motel, a traveling sales-man for one of a thousand products.

As I drove on, I thought of the pelican, soaring from mid-continent to the shore. It was quite a stunt, although the bird could not be publicly acclaimed for an achievement that was not of its own volition. No Lindbergh there, but the bird was a splendid voyager, all the same, and a great many people came to see it, so that it could have enjoyed some notoriety had it been the least bit interested. Other migrants were in my head, too, like the terns that would start to show up along the East-ern seaboard in a few weeks. They might already be on their way from South or Central America, going low over the water, with a quick, sure beat to their wings, veering from side to side with a searching motion fitted to the ocean airs.

Smokestacks released huge clouds into the air. Miles and miles of power lines marched ahead. I passed rivers that were opaque, like blinded eyes, and filthy streams, the gutted dregs of what were once pure, receptive veins in the life of earth. Tidelands were being dredged for real estate. Great areas of marshland had been destroyed by garbage dumps, landfill, or phosphate mining, tremendous stretches of the shore were torn up, stripped, and ravaged for their resources of shell, gravel, and sand. Silt-covered marine habitats smothered the shellfish. Spilled oil had been incorporated into marsh sediments, bays, and inlets. In some areas there seemed to be little life left, only the semblance of a grimy beach facing oily waters.

It had become a conscious effort to find wildlife. It was not that all America had turned into Hackensack, New Jersey. You could find clean marshes, native birds, and the sea beyond if you looked, but it was as if they only existed on our terms. The engineers could will it that the Mississippi be confined to ten miles wide, as compared with its original, great course, because the needs of motorized river traffic took precedence over the river's mutual accommodation with the land. We could, in theory, dam the Grand Canyon; or reroute the water from Canada, if we risked the political consequences. The power that moved us only seemed to respect nature's give and take when faced with some absolute limit to its actions.

The revolutionary aspect of our society does not depend on our having sent George III packing back to the old country, but on its experimentalism. We follow invention, what seems to work for the time being. There is a heroic ignorance in us that insists on taking what it can from the earth, if only because taking has made us what we are today. A triumph over circumstance built this great traffic that can take us where we want to go, and where on the way every man can dine as well as a king.

In the famous dialogue between the Lord and Job, after Job has already endured so many tribulations that he has desired to die, the Lord, out of his whirlwind, puts a series of unanswerable questions to him. The poor man *has* to submit, although he is justly rewarded for it later on, when the voice, the unassailable bully, asks: "Where wast thou when I laid the foundations of the earth:

"Dost thou know the balancing of the clouds, the wondrous works of him which is perfect in knowledge?

"Hast thou with him spread out the sky, which is strong and as a molten looking glass?"

Job survived the divine terror of God, the dark nature of creation, through the power of his humility. To endure as he did, to grovel in the dust, to be stripped of everything, had the end result of raising him in stature. He and his God, man's God now, were redefined. But if that Old Testament God has been replaced by man, what then? If such overweening questions were asked of many who ride the technological, industrial heap, they might well answer, "Sure. We were there. We know that. We can do all those things," and then sit back, although in some unquiet, existential limbo.

I sat in my motel room, trucks and planes droning, snorting, and making the ground tremble outside—my own blood trembling from my daily run in the car—while I drank a carbonated drink, read a syndicated newspaper, and shut off the TV to see a giggling image disappear down its throat. The thunder I heard was not the thunder of Jehovah; nor was I terrified like Pascal before the reaches of infinite space. As a matter of fact, the night was hard to see. Those stars by means of which birds may find their way over great distances were partly lost to view because of artificial light, smoke, and haze.

Where I stopped along the shores of the Carolinas, the wrens and cardinals were singing loud, and warblers were flying through lobolly pines, cabbage palms and palmettos, magnolia and holly. On Bull's Island, north of Charleston, I saw a wild turkey, displaying before his mate in the spring sunlight. His bronze feathers glowed like a dull fire. Shoals of light shifted over his back, giving it a glossy iridescence. With his tail opened into a great fan, and his wings spread and fluffed out to the side, this old American strutted like a ship loaded with sails, moving through heavy seas.

Canoeing around a bend of the Waccamah River, I heard the hooting of a barred owl, and there it sat ahead of us in broad daylight, motionless in a tail, red-barked southern pine. That night I heard it again, through the sound of jet planes ripping vents in the sky, as it moved through the swamp bordering the river, from tree to tree. Owls have a deep sensate realm of their own, calling to each other and their surroundings, feeling the expanding and contracting circles of the dark.

In Georgia, a kingfisher dove out of a palmetto, splashed into lazily moving water, then flew up with a loud, rattling cry. It was one of those snaking, shifting, weaving rivers that go through brackish marshes where small fish leap like fired rockets between the grasses and then subside, while the long, gray twists of Spanish moss hang and rustle from the live oaks on the banks. A big pileated woodpecker hitched up the bark of a pine, an erect, black body with a flaming, scarlet crest, giving a raucous, clarion shout. They called it the "Lord God bird" in parts of the South. Warblers lightly buzzing and trilling veered in and out of tree trunks, catching insects, tied to each other by their calls and the threads of their quickness. A common egret, with classic white wings, black legs stretched behind, rowed above the treeline. Orange and black monarch butterflies alternately flitted and made short glides over open ground; and a wood duck I had surprised at a bend in the river stared at me with an untapped lightning in its eyes.

In many areas, where sights like this are not just a normal part of your surroundings, they have to be found in wildlife or conservation refuges. We come off our omnipotent highways and cities to parks and natural areas where we hope to "see nature" and enjoy ourselves. They are places, according to the great Dr. Freud, which belong to "the mental realm of phantasy. A nature reserve preserves its original state . . . where everything, including what is useless and even what is noxious, can grow and proliferate as it pleases."*

Freud was clearly more learned on the subject of human fantasy than in the realities of the natural world, which, either at a distance or close at hand, are beginning to seem more basic to the needs of a whole earth than the ends toward which we try to steer it. Those who are concerned with endangered species, and trying to decide which of them to save first, do not get much help from adjectives like *noxious* or *useless*. They face emergency questions about living things that are being isolated

*Sigmund Freud, Introductory Lectures on Psycho-Analysis, In Strachey, J., trans. and ed., *The Complete Psychological Works*, Part 3 (New York: Norton, 1976).

from their sustenance. How can they be evaluated, how can they exist, without the original, full range of life in which all divisions, all contrary elements, are included and reconciled?

Brown pelicans flew along the coastal beaches of the South, gliding effortlessly together along the surfline, but they had been much reduced in numbers at that time because of their vulnerability to DDT and its derivatives. Cutting across to Louisiana, I found that breeding population of these birds had dropped to near zero. The mighty Mississippi, constantly pouring vast quantities of chemical wastes and pesticides into the gulf, had poisoned their food supply and permeated their tissues, so that they had raised no young. They and the ospreys, the bald eagles and other fish eaters so affected have been coming back from the brink in recent years because of restrictions on the use of DDT—which has not stopped our colossal and widespread use, often untested, of chemicals.

Many of the competitive and "harmful" insects have become resistant to pesticides, as if obeying prior directions as to survival. Other populations come back because they still have the space, the food supply, and the regenerative powers of nature behind them, but we do not often know the reasons for it. What do we need to know? What are the right equations between life and space, attrition and abundance? It often comes as a terrible surprise when we inflict so much damage on living things that it becomes irretrievable. What could we "lords of creation" have possibly done? A good guess is that we have been neglecting creation. It is certainly very hard to talk about man's responsibility for nature when we insist on being oblivious to it.

None of us know enough about that awkward-looking bird the pelican. Still, we might find out, somewhere along the line, that it was a linchpin in the wheel of survival that includes the human race.

At the Corkscrew Swamp Wildlife Sanctuary, on the north

edge of the Everglades, I heard a man cry out: "But, it's all so bloody alive!" as he looked at the teeming concentration of water plants, trees, birds, insects, lizards, frogs, and snakes. I might have been tempted to ask him whether it was so dead where he came from, except that I understood, having come in from some conquered territories, what he was talking about.

A swamp, from the point of view of people who "don't like nature," and there many, is a sink hole made up of blind, mutual consumption, and the most direct equivalent of it in ourselves is a darkness with no promises; but it can also be a source of relief and joy, as we discover that the earth is richer than our compulsions make it.

It amazed me that a huge area like the Everglades, once synonomous with remoteness, the nearly unreachable wild, should have been so cut down, not only in reality but concept. The pressures were obvious, from the canals to the north that controlled the flow of that wide, slow-moving river toward the gulf, to a threatened jetport and a future city of a million people. Real estate development, the multimillion-dollar business of recreation, with marsh-skimming boats and other destroyers of remoteness, just the constant clamor of a growing population demanding to be let in everywhere, made me thankful for the national park. But the human tide is relentless and conditions us all. We were seeing wild America, if we saw it at all, not in terms of its supreme diversity and running skies but of the outline, the bare bones exploitation could make of it, more real than its own substance.

I emerged at the end of this trip to walk down a west Florida beach, in a resort area. As I was looking out over the marl-clouded waters, I heard a voice, in low desperate tones, saying: "I don't know what to do." So I walked over to a man I had only half noticed where he stood with a fishing rod on a concrete wall, and what should he be trying to reel in, like a great stone, but a white pelican, which had the lure and line slung around its wing. As he kept reeling in, the bird let itself be pulled in for a foot or so at a time, a dead weight in the water, then flapped up again; and so it went, foot by foot, or inch by inch; each short haul bringing it a little closer to the dock. Then a second pelican flew in and dropped down close to the other.

"Keep off, you," the man growled. He had graying, close-cropped hair, gaps in his teeth, eyes slightly screwed up, and a broken nose. He was a slim, slight, but rangy man, his limbs shaky from the ordeal—perhaps they were always a little shaky—and he was quiet and apologetic in a manner that seemed part of his nature, too. I liked him right away, and while he reeled in I waited, lying down on my stomach, for the docking of the pelican, which was finally pulled in close enough for me to reach.

"Take a hold of his bill," he said. "Not that it hurts much, but it stops him."

The bird's combined bill and pouch surprised me when I grabbed it, being as soft as chamois. Its white eyes were pink-rimmed and stared vacuously, and after the line was unwound it flapped non-committally away—Lindbergh altered into Pagliacci.

The drone of engines; the grinding of gears; shore mud that smelled of gas; the ebb and flow of cash in voices everywhere; some loud kids and a quiet heron. The heron felt the sandy mud, its long, jointed legs and flexible, clawed feet lifting very slowly and deliberately. Stalking is the way to sense the ground beneath you, a delicate pacing, the steady feel of counterresponse in an incongruously long-winged frame. To find your food, you trace your way, a slave of patience from our point of view, but wise enough.

I saw a little migrant warbler come flying in off the windy waters of the gulf to land on the ground at the foot of a tree, looking genuinely distressed and tired, its feathers wet and ruffled. I listened to noisy willets where the tidal currents ran by. They had a near idiotic, brassy cry, flung off into the wind; while two hunter dolphins, wet flanks slip-streaming, arched in the rip tide, running together, their fins in a spliced play. A flock of white terns flew loose and easy over the water, dropping and diving as they went. A plover flew off and away with a sharp, sea-honed cry: "Keealee!" I watched the mating of horseshoe crabs on the beach at the high tide line, as the water was starting to fall back. Some got flipped over by the surf and had to wait for it to right them again. They were a gray-green color, unlike their brown relatives in the north, and their shells were encrusted with barnacles and slipper shells. Nosing into the sand, with improbable compound eyes like windows encased in martial helmets, they shoved ahead and into the sand, clasping each other, hitching rides-slow, cold, deliberate sex, out of ageless law. They stayed at the high tide line while the surf still washed it and then began gradually to move back out into the water, being fair swimmers once they reached it, and disappeared into a Cambrian depth.

One afternoon, a south wind was roughing up the cloudy green waters and rattling the palm branches. Thunder boomed; low and heavy, it spread around. A grackle ducked suddenly under a bolt of lightning, and so did I. Then the rain slammed and spattered down, while vultures, terns, laughing gulls, ospreys, and herons flew just inshore to a mile or so farther back, lifting and dropping down to the measure of the receding storm. The terns ran before the wind like leaves in a northern fall and occasionally shook themselves in flight during the hard pelting of the rain.

As the dark clouds blew off and the sun came through, the birds kept circling overhead, riding the updrafts, drying their feathers and evidently enjoying themselves. The clearing off resulted in the most subtle color changes in the sky and over the water. There were dark green and copper openings through the clouds hanging above the shore, clouds of violet reflecting the sunlight, others keeping a darker gray from the storm. The elements had exploded like a wave losing its energy in falling on the beach, and every life responded.

I met my friend of the pelican incident as we were plodding down the shore, drying out. "Couldn't get away from *that*, could we?" said he.

While development was doing its best to conceal the natural state of Florida, the exercise of light was never stopped for a moment. I looked to its participants to see how it was done. Freud's fantasy lands were exiled from earth experience. Travels are universal, so the white pelican told me, or they have no meaning.

CHAPTER 11

Space Craft

Wherever I have been, during the course of my intermittent travels, I have appropriated symbols of the
places I visited. They have not necesarily been as common or
visible as the birds and flowers chosen to represent the states,
such as the lupin or "bluebonnet" for Texas, the roadrunner
for New Mexico, or the cardinal that takes care of the seven
states of Illinois, Indiana, Kentucky, North Carolina, Ohio,
Virginia, and West Virginia, making a bright red slash across
east central North America. Some of them have only lit up my
horizon for an instant, like some spring warbler whisking
through a thicket. I have arbitrarily chosen them from an intuitive sense of what was appropriate, and perhaps essential, to
an area. They have momentarily enlightened me as to its vital
nature, filled up a missing space in me, told me that where I
landed wore a very special light, some of which might rub off
on me.

I even have some possible candidates I have never seen, such as the socorro isoped of New Mexico, or the yellow-nosed vole of Maine. They hover on the endangered species list, and since I have come to believe that the new phenomenon of accelerated extinction reflects on our own ability to exist in some kind of creative balance with the rest of life, I wonder what we may be losing when some of these lesser known orbits fade out.

The socorro isopods look like one of those sow bugs, "armadillos," or pill bugs, because of the way they can roll themselves up into a sphere when touched. Also called wood lice, they thrive in dampness. In fact, they require it because of their delicate gill-like breathing organs, and so can be found under rocks, stones, fallen logs, and rotting boards. A number of kinds live in fresh water, often creeping over muddy bottoms, but the majority are marine, found along the seashore between the tidelines. New Mexico was once part of a great inland sea. As its waters receded, the socorro managed to make the transition between salt water and fresh and survived in thermal springs, where it managed very well until we began to destroy its habitats. A remnant colony of these little creatures was found living in a drainage ditch by Mike Hatch of the New Mexico Fish and Game Department, who submitted them as a prospect for the endangered species list in 1976. At first, everyone treated it as a joke. "Bugs" have never been easy to accept, perhaps because they have prevented the human race from destroying itself through lack of the kind of competition that could hold it down. Isopods, of course, are not insects but arthopods, related to crabs and spiders, a plentiful, inoffensive, and varied race on the whole, but bugs will always be bugs.

Mike Hatch said:

> I have a hard time getting people interested in saving them. You know how game wardens are. If I told them there was a rogue

isopod loose down here, throwing boulders on the highway, they'd come in a flash. I'm always surprised by the limits of man's affection for the lower forms of life. Everyone likes puppies and kittens and baby deer, but that's about as far as it goes.*

Mike Hatch enjoyed watching isopods engage in "leaf surfing," climbing on leaves that slowly sank through the water and then getting off, letting the leaf rise to the surface, and climbing back on again for another ride. They would also glide through the water upside down, showing a white stripe on their bellies, and like Olympic swimmers they would come in for head-first landings, and then flip over. Up to now, they have been survivors of one hundred thirty million years. It is a skill not to be lost sight of.

Since it is typical of us to look down on lesser beings, it also caused considerable amusement among the legislators of Maine some years ago when the rare yellow-nosed vole was brought up in connection with its threatened habitat. This three-inch mouse lives in the higher elevations of the Sugarloaf Mountains, in cool, damp crevices, an area then being eyed by promoters of more chair lifts and access trails. Who could care about a mouse, especially one with a name like that? Said Dr. Robert Martin of the University of Maine at Farmington, who had studied this animal:

Few people, not knowing of yellow-nosed voles as part of their lives, would mourn the loss of this innocuous species, distinguished for its retiring habits. But it is difficult for me to consider man's voluntarily wishing to annihilate yet another in the long list of species he has destroyed on this earth, especially when it costs him so little to allow its survival.†

*As quoted in an interview with Molly Ivins, for the *New York Times*, January 1979.
†From an article by Phyllis Austin in the Maine *Times* of July 4, 1975.

There are many economic and utilitarian arguments on behalf of the need to save species. They can add to the texture and variety of the landscape; they are important to the stability of ecosystems. Research into the physiology of other animals gives science further insight into problems of human disease. There are many current and potential cures to be found in ocean life, such as compounds in sponges to cure encephalitis, agents in seaweed that inhibit growth of various types of infection, oil in menhaden that can treat atherosclerosis, and, not surprisingly, qualities in the nervous system of a squid that are of great value to research in neuroscience. Genetic engineering of plants helps in modernized agriculture, developing disease-resistant strains. New benefits for human health in the chemistry of plants are being discovered every day. In other words, we can stay healthy, live longer, perhaps forever, without ever needing to ask why. The system provides its own answers.

Our arguments on behalf of other species of life bow to our practical self-interest. They take it for granted that no new ethic to include them is possible without putting the economic, or even intellectual, benefits first. "What's in it for me?" or "What good is it?" has to be answered before we are going to be willing to move over to the other side. All of which still leaves us in the position of having an imperialist relationship to living things, giving ourselves the right to choose between one form and another. On a less discriminating level, it suggests that they are no more than items on a world shopping list. Utilitarians need only consult their own authorities on these matters, as you would consult your broker about investments. How are mice or isopods to justify themselves? Only through our willingness to let them in and be enlarged by their existence.

Since my knowledge of the state of New Jersey used to be limited to the Jersey Turnpike on the way to Washington, D.C., I was surprised to learn of a pine barrens tree frog there that was endangered because of the rapid development of its habitat by housing and industry. How vital was it in the scheme of things, either as an intimate of the mastadons or as a key player in that wheel of change which is, and was, New Jersey? Would we, if we lost it, have thrown away part of the map of the continent?

Should I disregard the news that the tiger beetles of Southern California have been almost totally destroyed because of oil spills, urban expansion, recreational use, and pesticides? They are quick and agile flyers and can run quickly, but they are not fast enough to get away from us. We in turn are losing the beaches, so it would seem that the presence of tiger beetles might indicate that the sands we were lying on to enjoy the sun were relatively safe and clean. That is one of those practical arguments we seem to require. In a different sense, I suspect that a tiger beetle might be as essential to the life of the beach as we are, probably more so, since it is not in a position to destroy it. Do not neglect your companions. They may lead you toward a better world.

Roy Bedichek, the Texas naturalist, with whom I only had a brief acquaintance, was, all the same, one of the best storytellers I ever met. He was rare among men in his humanity and intellectual interests—he started off his mornings by reading Plato—and he also typified the best his sunny, open land could produce. Like his friends Walter Prescott Webb, the historian, and Frank Dobie, teacher and writer of books about the West, who was a legend in Texas, he lived for the freedom, the great inviting openness of the Southwest. I once asked Walter Webb what characterized the region of West Texas the best, and he answered: "The capacity to waste time," by which he meant that they knew how not to let time ride their backs. It also implied that they had the room, and the freedom to choose from it, like the various natural inhabitants of their seemingly sparse range.

There is a warbler called the golden-cheeked, of the Edwards Plateau, in Texas, which has become endangered because rapid development has drastically reduced its habitat of ash juniper and the lacy-leaved Texas oak. This bird uses strips of the juniper bark for its nest and is particularly fond of a grub worm attracted to the oak. Because of this fairly narrow range of specialization, its is highly vulnerable.

Edgar Kincaid, the ornithologist, of Austin, was once taken out by Bedichek to look for the golden-cheeked, which had started out to go down in numbers, apparently because of collectors and cedar cutters.

> He took me out to a cedar brake that was alive with them. Thin clouds raced by that early morning so that the sun seemed a great ball rolling along just above the horizon. In memory it seems that almost every other cedar was crowned by a singing male, each with his brilliant black, white, and golden plumage in the perfection of early morning and early spring.*

Roy Bedichek and Frank Dobie used to sit on a rock, which they called Conversation Rock, in the waters of Barton Springs, the public swimming hole out on the edge of the city of Austin, where they would exchange characteristically pungent comments on life, literature, and the state of the nation. Bedichek called Barton Springs a "poem," with "great towering pecan trees over sparkling waters," and he described the whole south bank of the creek that fed the pool as being one solid kitchen midden built up through ages of occupation by Apaches and their predecessors. In a lighthearted letter to his friend John Henry Faulk, he expressed the hope that he might be reincarnated there as an Apache, "living on pecans and deer meat, working the excellent flint, and chasing my fleet-footed love

*Taken, along with other quotations that follow, from Ronnie Dugger, *Three Men in Texas* (Austin University of Texas Press, 1967).

over cliffs and down creek beds." But perhaps the time is not yet ripe for his return.

That was a good many years ago, when the population of Austin may have been about ninety thousand. Now, it is nearly five hundred thousand, and trouble has come to Barton Springs. On its approaches, park land owned by the city protects the trees and thickets along the banks of the winding creek that feeds the springs; but the high banks on the other have been developed, right up to the edge. The ground is occupied by housing and condominiums of a hitched and overlapping style that look, from the level of the creek, like battlements. The pure water that arrives through aquifers all the way from North Dakota became polluted in recent years, and the authorities had to investigate, while the swimming was closed off.

Corporate giantism has also moved in over the headwaters of the creek. Crowning one hill that once looked over an undulating landscape is a massive new complex of department stores that look like penitentiary blocks and cover two hundred acres with concrete. Around and beyond them new highways circle in all directions, and new building climbs the hills, obliterating the interchange of sight and motion as it used to be between the Texas oak, the juniper, and the sky.

It has been discovered in recent years that several species of the many local butterflies were in danger of disappearing altogether. Some of them winter in the area of Barton Springs. Others are colonists, like the heliconius, or zebra butterfly, which moves in from subtropical areas, picking out the same pecan trees to roost in year after year. When cold weather starts in the autumn, the northern varieties go into hibernation, and so the handsome black and yellow zebras have more nectar to feed on and may be present as late as December.

Population of local butterflies have always fluctuated because of climatic conditions. When they declined it was not so much the fault of overeager collectors with butterfly nets, nor could it be attributed to insecticides and herbicides. The forces of nature were the chief cause, but it is the loss or alteration of habitat that is now a far greater threat than the often violent give and take of the seasons.

Some of the smaller varieties, such as hairstreaks, skippers, and blues, which like limestone glades, juniper woodlands, dry bluffs shaded by cliffs, or sunny chapparal, are now extremely scarce. It is not that the areas they choose to feed in have been entirely wiped out, but that they have been so modified as to negate their very particular sense of them. Male butterflies, seeking the right places to mate in, are space conscious. They are aware of the structure of a habitat; and when an area of shrubbery or low trees is cut down or greatly altered they will have lost their landmarks. The identity of a habitat is strengthened by its diversity, and its impoverishment further reduces the life that is drawn to it. When I first heard that butterflies were space conscious, I thought to myself that they, rather than we heavy modifiers, must be the ones who held the land together.

New studies of coevolution, the way in which plants and animals have mutually evolved, deal with a dazzling array of countersignalling and relationships. Vision, visual memory, keenly tuned senses in a bird or an insect, are not separable from the plant growth of the world around them but have developed in direct response to it. The leaf signals its shape to the butterfly. Color, as it is to us, is a sign of warning or attraction. Nothing stands still or in isolation. The motions and behavior of animals affects the growth, nature, and chemistry of the plants with which they associate.

Why are fiery red flowers attractive to migrant humming-birds, which feed on their nectar and pollinate them? Because they can see them from a distance. Passing rapidly from one food source to another, their attention is caught by that bright signal. Only random signals, which can serve little purpose, have been eliminated by natural selection. They have to be well developed, as from one signaler to another, in all the close and useful exchanges that are the essence of the earth's vitality. In this, although plants may not speak to you, they are as sensitive and complex in their ways, as finely tuned and sophisticated in their rhythmic behavior, as any animal. One world accommodates the other.

Heliconius butterflies not only return to the same roosting spots over extended periods of time, they are also highly regular in their visits to sources of food. Laurence E. Gilbert of the University of Texas found that they would show up in the areas where they found nectar and pollen in the poor light of early morning and in the evening. This showed that light conditions were best for them at these times and also suggested that they had a circadian (about once a day) memory rhythm, as do bees. Their ability to find their way through sight is highly developed. In fact, they have the broadest visual spectrum in the entire animal kingdom, which is not surprising, since butterflies have heads that are largely made up of compound eyes and optic nerves. The heliconius has an exceptionally large head. Even in other aspects of its makeup and behavior, it is highly complex. Gilbert also found that the female pupae of a number of groups of this species release pheromones to attract the males. Females also make highly audible squeaks from inside the chrysallis, where the males can sit and wait to mate with them when they emerge.

Butterflies are brilliantly functioning partners in the vital character of a region, through the acuteness and discrimination of their senses. Like us, they need open space and its life-giving detail in order to see, to fulfill themselves, and to carry out its advantages. Space comes down to a sense of the appropriate, seen in two elderly, straight, and honest Texans enjoying the exchange of thought and speech many years ago on Conversation Rock.

Some thirty miles beyond the city limits of Austin is an area of innumerable springs lying under a deceptively dry-looking surface with a growth of scrub, mesquite, and oak. Jamie Anderson of Austin took me there recently to look for armadillos, since I had never seen one before. We stopped; we waited; we listened. We walked slowly ahead under the clouds and sunlight and light wind that made a play of shadow and light everywhere around us, running, dancing rhythms on the open ground. Leaves shifted suddenly. A bird called quietly. The place was full of a quiet magic. We were hunters of a kind. A sapsucker dropped from the trunk of a tree and whirred away when we came near. A hawk, about to land, spotted us and flew off. Turkey vultures wheeled overhead. A flock of quail suddenly burst out ahead of us and were lost in glistening grasses. Walking down a dry stream bed full of stones, Jamie caught sight of the armadillo, up to then a mythical beast for me. With its gray, rounded, and plated shell and a pair of sharp-pointed ears, it hurried off quickly enough, with an oddly delicate, quaint sort of gait. We found that we had disturbed it in the middle of eating a lizard, which it had dug out of its hole. The half-eaten body had a skin that was blood-tinged and sparkled like a mineral-rich stone.

"Oh, to realize space!" wrote Walt Whitman, who had a butterfly printed on the pages of the early editions of *Leaves of Grass*. He was a special favorite of Roy Bedichek's. It is also worth remembering that the butterfly was the symbol of rebirth and immortality among the ancient Greeks.

CHAPTER 12

Rescue Mission

Many years ago, my wife and I tried to rescue and clean up some oiled sea birds, when we had very little idea of how to go about it. A common household detergent had been suggested to us, but for all we knew its effects might be lethal. We were in the intervention business on a small scale, in that we were trying to do something about a problem that mankind, the great intervener, had brought about. It turned out to be a lesson in limitations.

It was a time when a great deal of oil had been jettisoned off tankers on the sea lanes to the north, and ocean currents had carried it down the coast, causing slicks over offshore waters and depositing some of it on the beaches. We had razor-billed auks in the bathtub, eiders and loons in the kitchen sink, over a period of several weeks. We found that feeding them sardines out of a can did not work very well, so we went down to the shore for clams and mussels. After feeding a little auk, or dov-ekie, for about ten days, I released it in a tidal inlet, where it whirred off on its short wings for fifty yards or so, but then dropped down to the surface, being too weak to fly. Feeling completely helpless, I watched it disappear, carried out on an ebbing tide.

We had also cleaned up a female eider duck, which we had kept for a couple of weeks while she regained the natural oil in her feathers. Oiled sea birds lose the insulating properties of their feathers. A spot no bigger than a fifty cent piece will result in exposing them to the cold so that they die of pneu-monia, or they ingest it when preening their plumage and are poisoned. In any case, we had not done as well with our duck as we had hoped. I was horrified, after releasing her, to see that eider, the epitome of bouyancy, start to sink below the surface. Her head and neck stuck out like the periscope of a submarine as she struggled to keep afloat. So I had to jump into those cold December waters and carry her home to try again.

The family technology was crude, and our success rate was practically nil, since I was never sure any bird we returned to salt water would survive in its probably weakened state. Since that time, an increasing number of individuals and conserva-tion societies have been on the lookout for oiled birds, and the methods of cleaning them, although far from a sure thing, have been much improved. Yet the birds still die of this affliction, often unseen, far out on the waters.

Many years later I was present at another releasing cere-mony that only proved to me once again how little we seemed to know about the worlds of life in any but a distant sense. This time the birds were all of the auk family, common murres and razorbills. They had been badly oiled but rescued and painstakingly cleaned and cared for. They had been penned up and fed on sand eels and Maine herring at the Cape Cod

Museum of Natural History after being picked up from the beaches in February. Auks often mistake oil slicks for the dark patches on surface waters that show them where small schools of fish are running. This particular concentration may have resulted from the kind of minor spills that occur all the time. In any case, some helicopters took off in search of it but found nothing.

It was now early April, and it seemed like a perfect day to free the birds. The offshore waters of Cape Cod Bay, in the area of Wellfleet, were calm and blue, and the sun shone across them with the clear, widening light of early spring. While we onlookers cheered them on, the little black and white, upright, penguin-like birds, were let out of the boxes they had been carried in, and they scuttled down the sloping beach into the water, where they ducked in with every appearance of satisfaction, shaking their wings and swimming out.

The little group, there were thirty of them, had moved several hundred yards offshore when I had a sinking feeling: "Where were the gulls?" The whole landscape had been empty of them, or so we thought, but just at that moment several herring and black-back gulls moved in and started to attack the auks, now small dark objects off in the distance. They kept diving on them repeatedly so that they had to keep ducking under, which began to weary them quite quickly. They grew weaker and weaker under this assault, and one of them no longer had the strength to escape, so a big black-back killed and carried it off. The survivors began to leave the salt water, which ought to have been their refuge, heading for the beach, where they landed like half-drowned people from a shipwreck. Nineteen out of the original thirty were picked up and transferred later on to a protected salt pond on Martha's Vineyard, which they left at a time that suited them, possibly at night, to travel north to their breeding grounds.

That their first release was unsuccessful shows how inadequate even the most conscientious efforts may be. To put yourself in the position of animals in their world you have to know more than most of us are capable of knowing. The habitat is as complexly tuned and varied as the life that shares it. Perhaps the feathers of the murres and razorbills had been cuffed and displaced in handling, or when they were being transported, with the result that they had no time to preen and realign them before hitting the salt water. That would account for their being quickly soaked as the gulls kept diving on them. It is also possible that the contrast between idle weeks spent in a relatively warm and dry indoor pen, followed by sudden exposure to the cold sea, may have been a shock to their systems. The time of day, the place of release, the preparations, were not quite right. So a group that demanded on unity for its strength became weak and divided. The people who care, and want to help, always have to know more than they ever thought they needed to.

During the winter months, some species of pelagic birds move down from northern and arctic regions to feed offshore along the northeast coast. But not too many people are aware of their existence unless they are out in a boat. Dovekies, or little auks, are occasionally driven inland by storms. During the winter months, gannets can be seen wheeling and diving with magnificent ease into surface waters from high in the air. Eiders, red-breasted mergansers, and scoters ride the offshore swells, but the dovekies, murres, and razorbills are usually farther out and seldom seen. It may be symptomatic of the age of oil that the auks are now a little better known that they used to be. It is as if they came in out of a clean world of once prodigal resources to try and correct our myopic vision.

I am all too conscious, in an age that piles up information like slag heaps, of my ignorance. What do we really need to know in order to live in harmony with the earth and to respect its integrity? If the only environment we pay attention to is the one we alter to suit our convenience, then we lose our inherited sense of a multitude of environments. We become the outside force that produces major displacements and disorientation on the face of the planet. Our fellow animals are robbed of an essential margin of safety we are unable to perceive because we no longer live with them. They turn from dancers of supreme skill to drowners. Will we become so detached from our sources in the company of existence as to be unable to rescue ourselves?

CHAPTER 13

Risks and Benefits, or, Death of the Innocents

The indomitable oceans where the sea birds ride over mineral-blue waters are forever pounding at the land, with its low or cliffed shores facing the turbulent horizon. They are vast in the circulation of their currents, vast in their capacity to sustain existence. They have always seemed boundless, forever on the move, but now they carry signs of an excess that travels beyond them.

The spread of oil on global seas is as hard to control as the fires of a burning town that has only one fire engine, but for the most part we are unaware of it except when it smears our boats and beaches. Oil slicks are too far away, as a rule, to affect us. Major spills and their animal casualties are more obvious, although most of us can avoid the implications, just as we avoid exposure to the kind of marriage with the elements that means life and death to a bird.

In December 1984 a tanker broke up off the California coast south of San Francisco, spilling two hundred thousand gallons of lube oil into the ocean, and the wreck was expected to go on leaking for years. Several thousand murres that nested on the Fallarone Islands to the north were coated with oil, and about a thousand of them were taken to centers to be cleaned. In the same month, fifteen hundred birds were oiled as a result of a smaller spill in Puget Sound. Human lives are occasionally lost as a result of disasters involving tankers, but for the most part it is the birds that die—and begin to die out, although the records for that are hard to find.

A few days after Christmas 1976 an aging, badly handled tanker, the *Argo Merchant*, broke up off the shoals of Nantucket Island, leaking out seven and a half million gallons of heavy oil into the Atlantic. During the same month, the *Grand Zenith* sunk somewhere off Nova Scotia and was never found. Thirty-eight crewmen were lost with the ship, which contained eight million gallons of oil that must eventually leak out of the sunken hull. The name of the *Argo Merchant* became known to every schoolchild on the coast of New England. The helicopter flew off and began to photograph the progress of the oil slick as it drifted westward. The fishermen were concerned about potential damage to the spawning grounds off Georges Bank. The tourist industry worried about the economic damage if the oil reached the shore.

The possible effect on wildlife was discussed, although peripherally, in the media. The courts of inquiry started to grind into gear. Television was full of the story for many days. It was reported that fifteen thousand birds, primarily gulls, had been found dead or dying on the Nantucket shore; but many people felt that gulls were expendable, overpopulated to begin with. Then the auks, loons, and sea ducks began to appear, shivering on the beaches. They were not found in great numbers, but no one could tell how many more might have died farther out to sea. Wind and currents kept oil from coating the mainland beaches of Massachusetts, and at the same time kept the oiled birds offshore. They were usually found on Cape Cod beaches to the north of the spill after a southwest wind, not when the wind blew directly from the north. In other words, most birds weakened by the spill were able to beat their way to land only on a southerly wind. So a great many more must have been lost at sea than were picked up.

Here and there, dead gulls could be found lying on the beaches, their bodies heavily soaked with dark brown oil, and occasionally an eider or a red-throated loon. The rounded black and white bodies of murres and razorbills lay on offshore ice or half sunk in the sand.

Since the wind and the currents were, fortunately, in the wrong direction to land the *Argo Merchant*'s cargo on shore, it was assumed, and reported in the news, that the oil was carried away and dissipated in the ocean without harm to anything but birds. However, in one area ten miles from where the tanker sank, a threefold decrease in benthnic organisms was subsequently found to have occurred on the sea floor. When oil reaches bottom sediments it will stay there for ten or fifteen years and continue to have toxic effects on marine organisms.

It was also a fact that the slick covered thirteen thousand square miles and that much of it floated under the surface, where it was at least temporarily lethal to species of plankton, the basic food of the sea. It also may have had an effect on fish populations, since fish eggs and fish larvae are easily killed by low concentrations of oil.

One afternoon I met a little murre who was standing at the edge of the surfline along Cape Cod's outer beach. It looked like a little statuette, although its head was moving nervously back and forth. Since a murre's legs are placed so far back on its body that it stands not on its webbed feet but on the shanks of its legs, or tarsi, and since its narrow wings are adapted mainly for diving and swimming underwater, takeoffs are difficult from a level beach. I tried to rescue this one but was unable to catch it. It was close enough to the water to escape me, flapping its short wings and propelling itself along the sands until it reached the waves. A murre is perfectly equipped to exist in northern seas and all their merciless exactions. It is tough, thickly feathered, and has a throaty, growling voice, a marine accent of its own. But nothing in evolution ever prepared it for oil. That bird, a matchless equation with the seas it depended upon, had been displaced, turned out, to wait for death on the sands, alone in a world slum it never made.

The family of alcids, including common and thick-billed murres, puffins, guillemots, razor-billed auks, and dovekies, are true oceanic birds, breeding in the spring on rocky islands up into the high arctic on both coasts. On the Atlantic side, they nest as far south as the Gulf of St. Lawrence and winter in the open sea down to the latitude of New Jersey, feeding on small fish, shrimp, or phytoplankton, each with a different method of catching its food. Murres and razorbills may dive as deep as two hundred feet below the surface, "flying" underwater in disciplined hunts for their prey. They are a wonderfully sturdy and hardy race, as clean in their plumage as the dark waters of the sea and the traveling clouds.

They stream toward their breeding grounds in early spring, engaging in elaborate ceremonial flights over the water. A crowded colony will engage in fierce territorial disputes for nesting space on the narrow ledges, crevices, and crevasses where the eggs are laid. Approaching a fog-shrouded island off Newfoundland in a dory, with waves splashing up on its rocky sides, I once heard the guttural sounds of a multitude of them, joined with the higher-pitched cries of the kittiwakes, dainty gulls that occupy the steeper ledges; it made a gabbling roar that almost drowned out the sound of the surf.

In some far northern regions, murres and auks lay their eggs on thawing ice, which requires that they brood and protect their eggs with even greater care to keep them at the right temperature. A constant watchfulness is also needed in order to protect the chicks from predatory gulls and skuas. There are natural factors in the life of these birds that aid their survival. They usually lay only one egg, which means fewer mouths to feed. Despite a large number of eggs lost to storms or tumbled off ledges during disputes, plus the loss of chicks, the adult survival rate is high. Young birds take three years to mature, during which time they learn to live at sea and depend on its pelagic food, away from the whole ecology of planktonic organisms and fish that surrounds their islands. Their strong, competitive feelings on the nesting territories help individuals retain their sites, and the fact that these outer islands and their high cliffs are hard for land-based predators to reach also improves their chances.

Alcids are good to eat, as coastal dwelling people have known for thousands of years. Indians and Eskimos hunted murres, dovekies, and auks. Greenlanders and Newfoundland fishermen still do and have often depended on them for an extra source of food. Although local hunting pressure may have nearly wiped them out on a number of islands, enough managed to persist before their populations were reduced beyond their capacity to sustain themselves. Visiting whalers and other seamen from Europe killed many thousands of them in a single day and even used their meat for bait. This led to the extinction of the great auk, a flightless bird that could not escape. But the whalers left, human habits changed, the rocky islands were no longer visited, or were protected by law, and so these very hardy and handsome birds began to increase in numbers again, until our omniverous technology began to reach into every corner of the earth. The new hazard of oiling may be a more impersonal means of slaughter than used to be engaged in by a boatload of sailors, but it is no less real, and far more widespread. The fact that the alcids lay only one egg and that young birds are slow to reach maturity means that an outside, unnatural, and destructive factor such as an oil spill is much harder to recover from. Normally, they need fifty years to double their populations. So that if a colony were to be reduced by half as a result of such a spill it would take half a century for it to regain its numbers.

Enormous disgrace is periodically committed on the body of the earth. In March of 1978 the two hundred thousand–ton tanker *Amoco Cadiz* ran aground on the Britany coast of France. The vast quantity of oil leaking out of the wreck turned productive salt marshes into ghastly dead ranges, destroyed famous oyster beds, coated seaside rocks and algae, killed off worms, snails, limpets, amphipods, crabs, and fish. That year's class of flat fish was wiped out. Inshore, there was a fifty to eighty percent decrease in mullets, and offshore a fifty to sixty percent decrease. Other fish species suffered from rot and sores. Those spawned later on were much decreased in size. Crabs were also smaller than normal, and the extent to which the reproductive system in oysters and other marine animals had been affected was difficult to assess. So was the long-term damage to an ecosystem made up of a multiplicity and complex interchange that has never stopped for analysis. And what an insult to the self-respect of those people who loved their land!

Less massive spills have nearly destroyed local populations of terns at various times, all the way from islands off Europe to the coast of Magellan. Thousands of penguins have been coated with oil and died off the coast of Africa. Oil has killed off sea otters, seals, and sea lions. Gulls with no oil on their bodies have been found to have a substantial amount of fuel oil hydrocarbons in their muscle tissue and aromatic compounds in their brains.

There used to be some mystery about where baby sea turtles went after they left the beaches where they were hatched and entered predatory waters. Archie Carr, who has devoted much of his life to their conservation, now reports that they appear to survive during this crucial period in their lives by swimming out to sea and riding on rafts of sargassum weed, where they are relatively free of predators, although sharks, groupers, and other fish may find them, and frigate birds, gulls, or pelicans may pick them off. Life has always been highly dangerous for young turtles, but their mortality rate is now rising not because of natural enemies so much as the added factor of widespread pollutants. The currents that form these weed masses also bring in pieces of styrofoam and globs of tar and oil. The little turtles bite into it, their jaws get glued together, and they die.

Much of the oil pollution in coastal seas has to do with the convenience of driving a car, a personal dilemma I have not yet resolved. It has been estimated that accidents involving tankers only account for three percent of the world's total. Much of it comes from "industrial and municipal effluents, urban runoff, and river runoff carrying oil from inland areas. A substantial amount of oil, therefore, will be discharged to the coastal zone regardless of sources."*

In other words, most of this pollution is the result of the normal operations of our society. When the automobile first came in, made possible by an abundant and inexpensive source of fuel, it was a delight to be able to drive anywhere you wanted to, although your rate of speed was only such as to scare the daylights out of a horse, or a flock of chickens. But machines, surpassing the restrictions nature put in the way of profligacy, began to gain more momentum than ever seemed possible. Two world wars increased the speed by quantum leaps, and the

*Harry M.: Ohlendorf, "Exposure of Marine Birds of Environmental Pollution," U.S. Fish and Wildlife Research Report, no. 9.

acceleration continues. We have become mesmerized by it. Increasing injury to the earth is justified in the name of economic growth. The temperature of the economy, our "life blood," is checked once a minute. If its ravenous wants are not being met, it falls ill, and so do we.

Modern industrial societies talk about tradeoffs between what they derive from the earth, loosely called "the environment," and what it suffers at their hands. They deal in "acceptable risks." In any drilling operation, and in the worldwide traffic of tankers, a certain amount of oil discharge is inevitable. Even when loading and unloading ships, there is some spillage. The accumulative effect of constant, routine leaks is largely unknown. At the same time, techniques of offshore drilling have been improved. More safeguards have been put into effect so that spills are less likely to occur; and new protective measures have been built into law. If I were to limit my reading to objective studies and the reports of industry, I think I could rest assured that we were doing everything we could to correct our mistakes. Still, petroleum and its by-products and derivatives continue to leak out across the globe and invade life's tissues. Is this acceptable?

How much is too much? Very often, as in the untested use of herbicides and pesticides, we have no idea until we see the results. We fling our waste products wholesale upon the land and water and justify the damage after the fact. The attrition the earth is suffering at our hands is excused in the name of all sorts of things, from human superiority over the rest of life to our absolute need to give ourselves what we want to have. Everyone gets in on the spoils, while the rest of life has to take care of itself; but it is becoming clear enough that we can no longer depend on the regenerative power of nature while undermining it at the same time. During an age when plants and animals are already dwindling at an alarming and accelerating rate as a result of the destruction of their habitats, how much more can they tolerate? How much more of a shove does it take to send any given population into an irreversible decline?

After the appalling accident at Bhopal, India, in December 1984, when some twenty-five hundred people were killed by an explosion at the chemical factory used to manufacture pesticides and two hundred thousand more were seriously injured, the "experts," whose allegiance to anything but the profit motive is hard to discern, were quoted as saying that the disaster was the result of unpredictable error, the weakness of individuals. But what erred in the first place except the weakness and ignorance of institutions? Since when, also, has the manufacture of death been a basis for sound judgment?

A professor of technology was quoted in the news as saying: "Of those people killed, half would not have been alive today if it weren't for that plant, and the modern health standards made possible by the wide use of pesticides." Pollutants and chemical wastes from similar factories and industrial plants have been destroying forests, fish, birds, and now, more obviously, people, all over the world. Is this to be described as health? That the continued destruction of life is being viewed in terms of a "risk–benefit" analysis shows a perversion of logic that is hard to believe. I guess, while I am driving down the highway, that I had better start an in-depth analysis of the extent to which the benefit has now become the risk.

Oil as we employ it is neutral. It has nothing to do with plants and animals but is directed by human will, quite outside their destinies. So we put ourselves outside of nature's protection, depending on technological fixes to solve our difficulties, or on simple arrogance. Can we improve on the Grand Canyon, or on life itself? Not without consulting them, never through elevating ourselves above that deeper sea which has the whole company of life in its keeping and is the only guarantee of permanence.

The sturdy bodies of the murres literally dive off their formidable, rocky cliffs and then whirr across the water like so many great bees. They were created out of marine tumult and power. The world they depend on for their sustenance is one of the stormiest on earth. They have to endure gale force winds, great waves that batter their nesting sites, storms of sleet or violent rain; but in the spring the upwelling waters around their nesting sites are wonderfully rich in nutrients, with an explosion of plankton that attracts an abundance of fish.

When I look out on the winter sea, I marvel at their endurance. I watch the waters from my protected distance. Ice rafts in along the shore after a week of very cold weather but then breaks up with warmer temperatures and disappears. Snow and sleet hiss over wind-drifted waves. Storms bring a darkness that is followed by a grand revelation of light, sun fires, and colors playing over the open plains of the sea. The murres and auks are its sacrificial victims, continually losing out to predation and accidents, yet beautifully made, a match for the elements that created them. We may imagine that their world is not our own, in the sense that we are no longer obligated to its uncompromising, wilderness laws, but this may be our greatest illusion. The source of life is the whole of life, eternally followed, through unfathomed sight.

CHAPTER 14

Flight of the Snapping Turtle

While I am still on the fast track, I think I should intro-
duce the subject of an animal that does not believe in
our way of life and has in the past served to temper my own
impatience. This member of the race of turtles, holding back
while we travel beyond all known limits, emerges now and
then from the Saurian depths of the neighborhood to cause
surprise and even consternation.

A huge turtle occupies the middle of the highway so as to
hold up traffic. The police are called in, but by the time they
arrive, sirens screaming, lights flashing, it has lumbered away,
thus avoiding much human commotion and possible disaster
to itself. Another such improbable beast arrives on someone's
front lawn, coming in, as it were, out of nowhere, or perhaps
from a television set. As a matter of fact, the snapping turtle,
Chelydra serpentina serpentina, being the species common to much

of eastern and central North America, is as primitive an animal as you are likely to find on these developed premises. It is descended from reptiles that existed several hundred million years ago.

Ordinarily, the snapper spends its life in fresh or brackish water, venturing on occasion into salt water. It often lies in wait along muddy banks or shallow bottoms, or simply floats in the water with its eyes just above the surface, not unlike an alligator. During the winter, although they are very hardy animals, those in colder climates hibernate beneath mud and debris, sometimes in muskrat holes. When the adults emerge in the spring, they may wander considerable distances, perhaps half a mile or more. The female lays between fourteen and thirty eggs with tough, white shells, digging a flask-shaped nest six to seven inches deep, covering it with dirt and leaving it. The young hatch out in the fall and head downhill toward water.

Snapping turtles are usually considered to be very ugly brutes, although I suppose aesthetic bias in these matters depends a good deal on what species you belong to. They are occasionally shot for quite arbitrary reasons, because people do not like their appearance, or are afraid of them, since they have a reputation for an ugly temper. For the most part they have been taken for their meat, which is one good reason for their conservation, at least from a commercial point of view. Years ago, a neighbor shot a number of them because he said they were eating our local alewives during the spring run from the sea, and since he had worked hard at building a ladder for these fish so that they could have easier access to their spawning grounds, he did not want to have his labor spent in vain. A year or so later I asked him whether there had been any effect on the ecology of the area, a small pond whose outlet led through the marshes to salt water, and he told me that the pond was now so thick with tadpoles that you could hardly see the bottom. So this might serve as another argument for the snapping turtle, although if it has carried enough life force to survive this long, that in itself might be enough for the defense.

Because of their persecution, the snappers seen these days do not often weigh much over thirty pounds, despite their reputation for great size (although one text refers to a specimen, fed on swill, that reached eighty-four pounds). They eat many kinds of fish, as well as crayfish, frogs, tadpoles, and carrion of various description. They also eat waterfowl that are unwary enough to come within reach of their iron jaws, which is not as often as people who are fond of ducklings might imagine. Young ducks are usually speedy enough to get out of the way. If not, as James Lazell, the biologist, has written, in a no-nonsense style, this is probably because they are sick and ill adapted "in otherwise healthy broods" and "become turtle food, as they deserve to. Turtle predation is thus beneficial to our native species."*

Snapping turtles have an exaggerated reputation for savagery. There was an old superstition that when their jaws closed on a victim they would not let go until there was a clap of thunder, or until the sun went down. All the same, a snapper only shows its rage when caught out in the open. They will sometimes rear up at you, or the stick you poke at them, with such sudden, impotent emotion as to lose their balance and fall over. When in the water, on the other hand, they react harmlessly, even pulling in their heads if you happen to step on their shells.

Evolution put the snapper's flat, armored shell high above its legs, so that it is able to walk on pond or river bottoms, but it swims poorly; and with its light-shy eyes it might well be angry when caught out in the glare of our world's publicity. It is infested with an absolute cornucopia, for the scientific con-

---

*James D. Lazell, Jr., *This Broken Archipelago* (New York: Quadrangle / Times Books), 1976.

noisseur, of internal parasites, occupying the intestines, the alimentary canal, the bladder, and the heart. Leeches inhabit the area around the eye sockets and under its limbs. It also has the misfortune, if you want to look at it that way, of having a peanut-sized brain by comparison with its size.

One April day, the twenty-fourth of the month, as I have it in my notes, I saw a big snapper in the salt marsh, moving through the grass. Its loose, pink flesh hung elephant-like below the dark shell, as the whole incongruous body eased itself very slowly forward. Each wide, clawed foot was methodically put down—the animal almost flowed—and was then picked up again. It turned its head and pointed snout when I advanced toward it, and I saw the glistening in its bumpy, periscopic eyes.

I wondered what it could be taking in. How did this primordial creature with the dark and awkward frame receive experience? There must have been a long exchange between the nerve ends and the brain. Decisions in that head could be eons in the making, despite that frustrated, spontaneous anger it was capable of. To escape me, it headed for a narrow ditch in the marsh that had been dug for mosquito control, and when it reached the edge it fell in sideways because of the top-heavy shell. When the snapper righted itself, it started moving down the muddy path of the ditch with the same tortured slowness. This was a matter-of-fact, no-answers kind of animal. "I am. I abide," would be all, and profound enough.

The snapping turtle is well known for the tenacity with which it holds on to life. It can survive being terribly wounded long after less primitive and hardy spirits would expire. This may be consistent with the long life spans and endurance of turtles in general, and part of a great sequence, including insects on the other end of the spectrum like the ephemerids, whose adult lives may be counted in hours. The possibility that green turtles are not sexually mature until they reach forty or fifty years provides another astonishing example of a great run-on inheritance that surpasses all our interpretations of it. If relative length of life for the range of species can be thought of in terms of the endless variety of ways they can adapt to environmental conditions, it is also part of the interplay of life and death, the potentiality of timelessness.

Once, as I was driving hard down a cross-state highway, I saw a dark body coming up in the center lane, and in a matter of seconds it took the shape of a snapping turtle. That the poor creature had managed to reach that point without getting hit seemed miraculous; and it had a waste of traffic lanes to cross before it could reach safety. The big turtle waited, flat on the asphalt, neck and head stretched out and forward, peering in blank bewilderment. I wondered for an instant whether I should not stop the car, jump out, and grab it by its studded tail so as to lug it by stages across the road; but I might have had to rescue myself and some other people in the process, so I sped on, accompanied and chased by my own machine-age tribe. Whether the risk would have been worth it is not something I can decide. I have my regrets for both of us in our varieties of helplessness.

Unlike us, sophisticates of a later time, the American Indians knew that turtles walk so slowly because they carry the whole weight of the world on their backs. There may be something in it.

A beautiful creation myth of the Iroquois has a female carrier of new life descending into a great cloud sea, enveloped by a dazzling ray of light. Since the light needed a resting place, many animal divers in the upper level of the sea tried to find it. All of them failed except the muskrat, who brought back a small portion of material in his paw. "But it is heavy," said he, "and will grow fast. Who will bear it?"

The turtle was willing to carry it on his back. So the *oeh-da*, which was to grow into the whole earth, was placed on his shell; and now the water birds, guided by the glow of the great light, bore the woman down to the turtle on their wings.

So, as the story goes, "Hah-hu-nah, the turtle, became the earth bearer. When he stirs, the seas rise in great waves, and when restless and violent, earthquakes yawn and devour."*

During the rip-roaring days of the pioneers, the river raftsmen boasted that they were "half wild horse and half cock-eyed alligator, and the rest . . . is crooked snags and red-hot snappin' turtle." Having put aside this crude way of comparing ourselves to the animals, we now approach them from the point of view of a society that is able, on the surface at least, to have its own way. We rate them according to their established market value. By these standards, the uglies do not always fare very well.

"What good is it?" I was once asked by a close and dear relative as we observed an alligator lying out on a sunny bank in Florida. Now I knew her to be an admirer of civilized artifacts, of music, written history, and cultivated flowers; but this strange, dark animal with its ridged, scored hide and yellow slits for eyes, simply did not fit her ideas of order. It was as much of an unwanted accessory as mice in the flower beds, moles in the lawn, or spiders in the living room. To be fair, she was also afraid of it. We have not lost our fear of the wild and the unfamiliar, which may very well encourage unfamiliarity. We see a symbolic landscape, ordered by our ability to reproduce it. We also share an inner fear of the dangerous grounds on which our imperial order is based.

"What good is it?" The basic data is not yet in. At least we have slow time to count on. "Patience," says the turtle, "and you may yet learn the meaning of the good."

*Taken from Frederick W. Turner III, ed. *North American Indian Reader* (New York: Viking), 1974.

CHAPTER 15

The Immortal Present

*O*ur pursuit of time, aided by advanced technology, is less a matter of pursuing than pinching, as if we wanted to grasp the limits of the tangible. The accuracy of calculating a second now comes close to blowing up the world. This may be behind our denial of death, or treatment of it as an absentee phenomenon. Just as we have left innumerable other landmarks behind us, death now lacks the ceremony of place, once inseparable from an eternally recurring past. In a sense there can be no "passage of time" in a culture that rejects continuity. We model. We manage systems. We subject all things to endless analysis, while hunting for money is our basic game. We inevitably fantasize the world we live in.

The realities of earth existence have to be reconstructed, the way archeology delves into the "dead past," to try and bring it back to life again. This is exacting, patient work. The data are at hand, the context remote. Many layers of material, accumulated over thousands of years, may have to be sifted through. The possibilities of interpretation are endless, although the cycles of earth repeat themselves and human behavior has its consistencies. At least we can recognize the community we have always belonged to, no matter when its period in history may have been. The ground abides that we desert for outer space.

One afternoon in early November I accompanied an ecologist friend of mine on a visit to a New Hampshire bog. He was out to probe its depths by means of a piston corer, a long rod containing a cylinder that could extract samples of sediment, so as to test their age and chemical composition. The bog had mostly filled in a lake left by the glacial retreat twelve thousand years before; it extended out from banks covered by second growth trees, such as poplar, beech, sugar maple, red oak, and white pine. Off in the distance was Shaker Mountain, with birch trees showing like white whips in the green and purplish gray masses on its slopes.

These trees were migrants. Fourteen thousand years ago the land was open tundra, where nomadic tribes hunted musk ox and caribou. The oaks and white pines moved north into New England some eight thousand years ago. At sixty-five hundred years, the beech trees arrived. Hickory and chestnut came in late, about two thousand years ago, migrating from the Midwest, and probably delayed in their progress by mountain ranges running from north to south. Trees do not stand still, any more than the regions where they grow. They migrate in response to the earth's receding and advancing tides.

The township in which the bog was located was once a thriving community. A number of grist and saw mills were located along its principal stream, a long, fast-running brook; piles of rock were all that remained of them. This was now a sparsely settled area, nearly abandoned, except for the asphalt road, to whatever in-dwelling needs and cyclical business nature continued to attend to.

The ancient lake still survived as patches of open water beyond shrubby growth, reeds, and sedges. The marshy ground was full of hummocks and little islands where larch trees grew, now shedding their needles so as to make flaky, orange patches on the ground. Here and there were signs of beaver. They had chewed off branches of poplar and swamp maple and made clearings of mossy ground, like little lawns where they might stop in passing for a leisurely chat. Many trunks and branches stripped of their bark, as well as tunnels radiating through the grasses, gave the impression of a very busy and serious community.

The ecologist probed with his long rod, and at a depth of sixteen feet he extracted a core of "glacial flour," a silt-like, milky blue-gray substance that had been the result of the grinding action of the glacier on mountain rock. Much of it had probably been carried within the body of the glacier itself and then washed out by rains during the warming trend that accompanied and followed the retreat. Material gradually filled the bottom of the lake, and on top of that a big mat began to form, the result of a long accumulation of the fragments of sphagnum moss. So the cylinder, or "marl and peat sampler," came up with the vital evidence, shaped as if it had been accumulating in the earth's intestinal tract for thousands of years. The poet Walt Whitman, who wrote that he incorporated "gneiss and coal and long-threaded moss and fruits and grains" would probably have been happy to add glacial flour to his collection.

I watched the material being extracted, and to me, who had never dove much deeper than the old swimming hole, much less twelve thousand years, it was breathtaking. The aluminum rod came up tundra-cold. It was like disinterring the far end of things, bringing it up out of its icy hiding.

All things are measured by man's ability to interpret and to count beyond the immediate; at the same time, how can temporal experience count all the changes that surround it? By evening, the local weather began to respond to global, continental movement. Weather from the south, in a huge, swirling interchange of air masses, began to give way to weather from the west. Marvelous, brief showers drifted across the bog, completely wetting us, then holding back as if in response to some major hesitancy in the accumulating powers of the sky.

It was growing dark. A carload of hunters moved slowly down the road, sweeping it with their headlights, on the look-out for any deer that might be moving down from the surrounding hills. (Mechanization may help us to shoot our prey from an easier distance. Bringing "wildlife" closer to us in terms of a life and death alliance is another matter, yet to be resolved.) We climbed the banks below the road, after carrying the equipment through a thick growth of sedges and brushing past sweet gale, leatherleaf, bog laurel, and bog rosemary; those plants that define the ground, in particularity and endurance. They give it its northern face. How beautiful to be the leatherleaf, or cotton grass, with the gift to shape the landscape. The sphagnum moss that forms the spongy body of a bog is remarkable enough in its equation with water. Through numerous storage cells it is able to hold thirty times its own weight. Each to the miracle of participation.

The bog was undeveloped; it had survived being "unimproved." So all its life was free to sustain its essential nature, collaborating at the center of past, present, and future, under the carrying in and out, the endless passage of the wind. One plant knows a million or a thousand years; its roots are endowed with immeasurable possibilities. The one consistency is eternal motion in its form and place. And am I not a breathing, flesh and blood chronometer of the wind?

CHAPTER 16

Custodians of Space

The exploitation of trees has led us from rags to riches, and here we are suspended on our precarious height, regarding the developed ground where the great trees used to flourish and wondering how far we may have to back down. To be able to lay low the greatest deciduous forest on earth, once occupying a large part of this continent, was no mean feat. What subsequent inventions could ever exceed that ax-wielding test of progress!

New Hampshire, where I spent much of my boyhood, is tree country, and the white pine, which favors regions of moderate to heavy snowfall, is one of its dominant species. I remember one stand, bordering a wooded road, that had been left to grow and had reached a considerable height. The pines formed vaulted aisles like a Gothic cathedral. I looked up and they sang in my spirit as the wind swished through their long, fine needles swinging in the light.

The great hurricane of 1938 cut a swath through inland New England and hit our lakeside area head on. All those cathedral pines and others pluming by the lake shore for a hundred years or more went down like so many matchsticks. My father, who loved trees, came up from New York where he worked to view the damage, sat down by the road, and cried. But since this is a land of trees—in the character and quality of the light they are adapted to, and in the cold and stony soil which the old time farmers cleared and plowed—it is also irrepressible.

In some areas where white pines had grown in on abandoned fields and then been knocked down by the hurricane, they were succeeded by an understory of beech trees, with smooth, silvery gray trunks reminiscent of marine fish, or shining white birch, or sugar maples and ash. Following their pattern, the pines seeded into open clearings and unused pasture land, taking the opportunity to reclaim thousands of acres for themselves.

Either as farmland or original forest, these northern farmlands exacted their tribute from human beings, certainly their respect. They forced a hard life and cantankerous natures on the natives. The long, dark, snowy winters and the short growing season, which to trees are perfectly natural, are not easy on human dispositions. But at least the people were molded by the place they lived in. As Wendell Berry put it, in *The Unsettling of America:* "If we do not live where we work, and when we work, we are wasting our lives and our work, too."

One fall afternoon I was walking down the lower slopes of a minor mountain. It was toward sundown and, since it was now December, getting much colder. Behind me, a screech owl wailed in the shadow of a belt of brilliantly white birch that lay between mixed evergreens and hardwoods. Otherwise the air was still. I walked out into a clearing, where there had once been a small farm, with nothing left but a cellar hole and open fields before it, that was crowded with white pine seedlings. Below that, a long stone wall descended to a cold, clear, rocky brook perpetually sounding. I could feel the edges of an overload of freezing air that was about to fall and turn the grasses white. The ground, too, seemed withheld, waiting in silent strength.

The original house, I knew, had a sometimes grim understanding with the north. A combat had been waged here between life and its limits. People had been close-mouthed in the process. "Nothing to recommend," I heard them say. But they were centered in time and place, and there was a tall sky and long hills to look out on and remember. Give them credit for the way all seasons indentured them.

They had left their doorstep behind them, a great slab of granite, and a big sugar maple. Its broad-beamed trunk, covered with shaggy, deeply ridged, gray bark, stood over the cellar hold like a reliable ancestor and descendant. The field where they had grown corn and vegetables was returning to its original symmetry. The wilderness air had its way. A chorus of crows responded as cold, rushing waters fell toward the base of the hills. Soon, blue stars would shine out in the well of night. I listened to a silence that followed me away.

There are occasions when you can hear the mysterious language of the Earth, in water, or coming through the trees, emanating from the mosses, seeping through the undercurrents of the soil, but you have to be willing to wait and receive. And there is a planetary silence behind it that defines the unseen quality of existence, as on a day when the white pines—one of the founders of America—are loaded with snow and nothing stirs. They seem to say: "What more do you want to know?" What more, indeed, can we know?

There was a white pine of fair height standing on a knoll over a little ravine that cut the hillside slopes above the lake, an area where my father had introduced many species of rock plants between the granite boulders. It was in that tree that I built a platform from which to view the world along with the red squirrels. I inhaled its resinous scent. The wind blew through the branches, sounding like the sea, and what distant worlds I imagined there I can no longer remember, although they probably included the walls of Jericho and the plains of Araby. White pine wood, soft and clean and easy to cut, also sent me on other voyages. I built a houseboat of it, a flat-bottomed boat with uptilted bow and stern and a cabin. It was powered by a Johnson outboard, or a long oar when the motor failed, with which I drove it slowly forward, and from it I fished for black bass and explored all the inlets and corners of the eleven-mile-long lake.

Having the mind of a mole, I built tunnels in the hillside below my treehouse. I built trenches there, too, being warminded, and awaited attack. The fringe of woods along the water's edge sounded with the liquid note of thrushes in the twilight. The lake itself made music with its wavelets for much of the year, except when it was iced over. They slapped at the timbers of our boat dock, or at the rocks along the shore, and they lapped at the edge of the ice in early winter. At times a stiff wind sprang up to chop the lake's surface to pieces, while at others the waters were wide, limpid, and glassy. I watched thunderstorms come in from the direction of Blue Mountain to the north, and the gathering noble blue-blackness would suddenly roll in overhead and the rain splashed across the lake, which ran with corrugated ripples, while the gods let loose their thunder and their bolts of lightning. If I was caught out in the open and there was no time to reach the house, I would go in under the pine for shelter. It is an unfortunate man or woman who has never loved a tree.

The power of forests is in that wild darkness the white settlers tried to get rid of; it never meant annihilation but stability. Trees, it has to be said, are "savages." They grow sick, they suffer from abuse and man's polluting tides, they inevitably die, as we do, when their span of life is finished; they are also uncompromising primitives, as anyone trying to make his way through a dense spruce forest will understand. In us, too, is the savage richness we override. Without those deep foundations, we would not understand the earth and would die out from lack of nutrition. Although it may temporarily satisfy millions of people, manufacturing goods and services beyond their need, a civilization that treats the trees as if they were inert ciphers, part of the numbers game, abandons the depth that builds community and invites vast areas of local sadness and vacancy.

In southern Mexico there is a remnant tribe of Mayans called the Lacadones. Their lives were always intimately dependent on the great mahogany forests that had lasted since the Mayan civilization died; but in recent years most of the great trees, now exploitable commodities, have been cut down. In *The Last Lords of the Palenque*, the tribal leader Chan Kin, then in his eighties, speaks in this way:

> "What the people of the city do not realize," he says in a heavy voice, "is that the roots of all living things are tied together. When a mighty tree is felled, a star falls from the sky. Before one chops down a mahogany, one should ask permission of the guardian of the forest, and one should ask permission of the guardian of the stars. Hachäkyum made the trees, and he also made the stars, and he made them from the same sand and clay, ashes and lime. When the great trees are cut down, the rain ends, and the forest turns to weeds and grass. In El Real, six hours from here, which used to be forest before the trees were felled, the top soil erodes and disappears, the streams have dried up, and the corn that grows there is stunted and dry. All becomes dry, not only here but in the highland as well—not only in this heaven, but the higher heavens above. Such is the punishment of Hachäkyum. I know that soon we must all die—all of us, not only the *hach winik*.

There is too much cold in the world now, and it has worked its way into the hearts of all living creatures and down into the roots of the grass and the trees. But I am not afraid. What saddens me is that I must live to see the felling of the trees and the drying up of the forest, so that all the animals die, one after the other, and only the snakes live and thrive in the thickets."*

The more the forests are destroyed, the more we turn into separatists, strangers in our own home. We lose our companions. We lose our way, because an age where all things are expendable makes it increasingly hard to identify what we need, and for that, reason is not enough. You cannot follow trees if they are not in you, but only in your way.

I once set out on a camping trip along the Deerfield River in Vermont. We had made our way to the campsite through woods made darker by a coming storm. We forded stony streams, slipped over wet logs, and tramped over shallow duff at the river's edge, while the wind kept increasing in intensity. Along the route were a number of big maples, covered with the yellow leaves of autumn, that made me very conscious of their presence. They had been growing there for a long time, stubborn roots probing piles of schist and a stretching skin of soil, until they had attained an eminence, affecting the character of everything around them.

The wind grew wilder and the skies blacker. The waters of the lake we were to camp by were whipped and torn, and finally intermittent showers changed to full and unrelenting rain. We packed up the tents and trekked back in the evening over trails that had been turned into running streams. We skirted the river, deep, wide, and dangerous, running down between woodland banks with a drawing power of its own, full of torrents and torment; it seemed to be gathering the growing darkness into

*Victor Perera and Robert D. Bruce, *The Last Lords of the Palenque* (Little-Brown, 1982).

itself. The wind hit the maples with a wild fury, so that their leaf masses whirled like a fire in the sky. The great storm, like the trees themselves, heightened the expressive unity of the region, its endurance, its stress, its ceaseless cohabitation. I was conscious of the power of interaction there, of bright leaves sensing change in the measure of seasons, of dark leaves and decay, and of all the transmutations unknown to me but carried by the creatures in the soil, and manifested in strange forms like the slime molds that moved on the trees.

We define our confused and disorderly world in terms of extremes we are unable to reconcile: war and peace, order and disorder, health and disease, life and death. Forests, on the other hand, have always contained these opposites in the unity of their being. Their wounds and malignancies are gradually healed. Wildness takes such care of wildness that it must always be the earth's criterion of health.

Trees stand deep within a kind of knowing that surpasses human knowledge. We are running too fast to absorb it. To go so far beyond them is to lose the sense of a community that thrives on the unities of the world. One day I was climbing up Sunset Hill, which overlooks Lake Sunapee, where we lived. (Sunapee, with various other spelling, such as *soo-ni-pee*, was an Indian name for wild geese.) A light rain was falling, coming in with a southeast wind. Most of the leaves on the trees had changed to copper, bronze, and brownish yellow, or they had eddied off into the wind. A pearly gray mist stole between the trees and hung over outcroppings of granite. The wind would occasionally push little shreds of cloud across the trail. I passed through a gaunt, dark belt of spruce with splintered arms that had haunted me as a child. I had imagined wolves running through them. Now the trees were filled with a slow-moving, sea-gray atmosphere.

As the light rain fell through the clearings and was filtered through the trees, the whole region had been changed by the watery atmosphere. There were countless new adjustments, recognitions, and responses set in motion everywhere. The hills and its woodlands received all outer weather with inner calm, and much art of its own. I watched while a junco, slate-gray and white like rocks and clouds, whisked and darted around in the underbrush and then alighted on the low branch of a spruce. I saw a raindrop suspended from a waxy needle in front of me, as it was swelling and about to fall. Water drop and junco, instead of being two separate and separable phenomena, became simply and easily allied. I was quite ready to say that a bird was like a raindrop.

The less we are able to admit common feelings into our relationship with trees, the more impoverished we become; it must indicate a deforestation of the spirit. Strangely enough, their least understood qualities lie in the sensate natures they share with the rest of life. When I walk through cut over areas where pasture birch, young sugar maples or white pines are growing back, I sense that they have a will of their own, an ability to come back that is more than automatic. After all, they are providers. They nurture multiplicity, from root to crown. Just as every life that associates with trees must communicate in one form or another, so trees themselves seem to respond to each other. We have hardly started to explore our mutual chemistry. On a high and open night in the winter, all blazing with the laddered, climbing stars, it is not accidental that the branches of the trees should reach and gesture as they do, or that one's spine should tingle at the lineup of the constellations. We were both constructed to that end.

It is December again, and the snow filters down through the air, while the wind picks up. Grainy-trunked ash, white and pinkish birch, dark green hemlock, ironwood trees with crinkly bark, all stand together in the snow laden wind. The higher branches creak and crack as they scrape or butt each other. They make a sound now and then like the tapping of a woodpecker. One tree suddenly sings like a bird, a singing note that is kindred to the wind, a sound that moves with air and snow, as we ourselves have voices that move in harmony at times with some deep, far off foundation for sound.

These trees might tell us where to whet our minds and appetites and examine our credentials for sight and hearing. They are keepers, not just of a wilderness apart, a reserve for the benefit of study and ecological research, but of a testing ground. A real acquaintance with them is more than good enough for the grace that living asks of us.

So in that uncompromising beauty of the arctic cold, you can walk where they stand on the hill tops and watch the wonderful sweeping by of the snow like smoke, with its visible particles bouncing, racing, intercirculating, on and on, with immortal energy.

CHAPTER 17

Open to the Sun

I first went to Costa Rica because of that small country's reputation for conserving its threatened resources. I also wanted
to follow out some life lines I had learned to look for at home,
such as a tanager, an oriole, or a swallow back for the nesting
season from the Southern Hemisphere. They do not conquer
distance so much as embody it. They are more beautifully tangible than any map; and when I see them fly in I have some
assurance that the rain forests are still intact, When they fail
to appear I worry about their future and our own. Birds are
indicators not only of the state of the lands they come from but
of how it goes with us who despoil them. Are we becoming
accustomed to emptiness? Are we leaving all the trees behind
us?

Exploitation was in the air as soon as I arrived at the airport
in San Jose, where I overheard an outgoing passenger, a fellow

American, telling how a friend of his had made a huge profit out of an unprotected stand of balsa wood. This was largely a matter of buying cheap, selling dear, and taking advantage of the greatly devalued colone, the currency in that nearly bankrupt country, then several billions dollars in debt. To his credit, he looked a little embarrassed as he reported on this shameless operation, but unembarrassed speculation goes on as usual, destroying forests and degrading soil, a process that has brought some countries to a kind of ecological death. Millions of people can no longer find enough sustenance in their own soil. Through its system of national parks, Costa Rica has taken steps to save some of its heritage before it is too late. Even so, it is said that at the rate the forests are going, all unprotected stands will disappear in a few years, a devastating prospect for a country that can boast two thousand known kinds of trees.

Tourists like me can skim like dragonflies between one resort area and another, although resorts seem relatively scarce in Costa Rica, or to those parks open to visitors. Then we clear out in a few weeks, leaving the country to its fruit and coffee plantations, its hundreds of thousands of cattle, its agriculture and expanding cities. You can soon see what population, industrial advance, and the exploitation of land is doing to try the country's resources to the limit, down to the last poor landholder facing the prospect of an acre of soil that has been finally drained of its life. But Costa Rica, unlike many other "Third World" countries, has been learning how to hold its own.

In the developed countries, through power acquired over nature, although derived from nature, millions can be deceived into thinking that the land is only a background for their material needs, and to be manipulated accordingly. For the developing countries, on the other hand, where development is still an unattained goal, or whose economic resources are limited, land is life and death, nutrition or starvation, expectations or burned out hopes. There is no way to disentangle local destinies from it. The land dies and the people die with it, often through social disruption and assaults on each other. The symbol for a country that has lost its resources in the land is a gun.

Costa Rica, to its everlasting credit, has no standing army and has spent much of its income on health and education. It is essentially a nation of small landowners and has a reputation for stability, although its economy, like most others in the developing world, is dangerously unbalanced. It is a country devoted to the needs of agriculture, while it has become increasingly dependent on the industrialized world outside it. The tropical forests, which once occupied 99.8 percent of its land area, have been reduced by nearly 70 percent. More than half of this deforestation has occurred since 1959, because of the demand for land, a growing population, economic pressures, serious invasion by squatters, exploitation by cattle and lumber interests. Even modern timber companies employing professional foresters have been involved in the loss of the rain forests, in Asia as well as Latin America. They have often been less interested in the practice of sustained forestry than in securing their own investments.

Cattle ranching, in which many Costa Ricans take pride, has also contributed to the loss, not only of forests but of the grasslands that have replaced them. When this land gives out, particularly on the hilly slopes that comprise much of the country, the cattle ranchers have often let in local people, the *campesinos*, to practice what agriculture they can. The result, in a few years, is a soil worth nothing at all. Under optimum conditions in the tropics it might take two centuries to regenerate. The average Costa Rican eats rice and beans, while those of us who frequent fast food eateries eat the hamburger, luncheon meat and frankfurters. It may seem like an innocent American rite, but in effect our exaggerated demand for meat, not only for ourselves but our canine pets, has encouraged the extinction of species through their lack of habitats.

Costa Rica is a small nation, no bigger than the state of West Virginia, and all it has left of that dark and intricate forest growth that extended from Central America to the Amazon, that magnificent original, are representative samples, although they are rich enough. For an outsider, a species count of the tropical rain forest does not mean a great deal until you encounter the rooted reality; but the numbers, with all they imply about what most of us do not know about living forms, are startling. When I visited Corcovado, Costa Rica's first national park, I was told that of the four hundred kinds of trees that scientists had found there, a possible one hundred more were still unknown. In Costa Rica as a whole, eight hundred forty-eight bird species have been found, twelve thousand flowering plants, one thousand ferns and their allies, fifteen hundred kinds of fungi, an astonishing variety of insect life, not fully explored. What will the disappearance of the world's rain forests mean, besides more deserts and disastrous changes in the atmosphere, but the loss of a balanced abundance we cannot ultimately do without? There lies in those tall, vine-covered trees, the light-loving butterflies, the stirring soil, the intensity of birds, all multiply engaged, that wilderness power we denied too soon at home.

Costa Rica has already lost an enormous amount of soil, the guarantee of any country's vitality. According to the Tropical Science Center of San Jose,

> 17 percent of the country is severely eroded and 24 percent is moderately eroded. . . . Serious and widespread erosion (30 percent of the area) occurs on the Pacific side with another 30 percent moderately eroded. A rough estimate of soil loss is 680 million tons a year to erosion, of which over 80 percent is caused by overgrazed pasture lands. Since topsoil is essentially a nonrenewable resource, the soil loss seriously threatens not only the country's agricultural productivity, but the economic viability of

hydroelectric, potable water, irrigation and forestry projects as well.*

It is a country defined by a great spinal column of mountains and a central valley where agriculture is practiced. The tilled soils also climb the slopes that characterize so much of the country. There you can see the gullies, a deep clawing into the ground, the giving way of an underlying balance, as well as brown fields terraced by wandering cows, or red earth marked by blackened stumps as memorials to the vanished trees. In some areas black weathered rocks give an additionally blasted look to the landscape. The traveler sees fires as he looks down from a plane, fires from the roads and highways. In March, during the dry season in Guanacaste Province, it looked as if the land were burning everywhere. Fires climbed the hillsides; smoke filled the air and darkened the sky. The pastures, once forest lands and already overgrazed, were being set on fire in the annual practice of burning off to prepare them for planting. How many years the soil, open to the fierce tropical sun and to periodically heavy rains, would hold out was problematical.

A rain forest has a fire of a different kind, annealing and regenerating, perpetuating those planetary energies that are fueled by the sun. The forest reaches upward toward the light. The trees are not sheltered from it, they employ and embody it, and every life they shelter shares in that exchange. The rain forests have a stability that has been millions of years in the making. The annual rainfall in Costa Rica comes to at least four meters and is not a limiting factor in the growth of the plants, nor is the temperature, balanced within the system. The principal, dynamic factor affecting forest growth is light. Forest nutrients are locked up in the vegetation and only superficially penetrate into the subsoil. The topsoil has only a

*From *Country Environmental Profile: A Field Study* (San Jose, Costa Rica: Tropical Science Center), 1982.

thin layer of humus as compared with temperate zones. The leaves continually drop down to the forest floor, instead of seasonally, and they decay fairly rapidly. Nutrients are taken up by a tightly linked network of roots and associated fungi and transmitted to the growth of the trees.

Fire the rain forest, cut it down, break open the soil to the unshielded light of the sun, and its life forms are wiped out, or forced to live on a marginal subsistence level in whatever pockets of sustaining habitats remain. Every acre so destroyed can lose literally thousands of plants and animals. On the global scale at which this occurs, an instability results, of a peculiarly modern kind, the consequence of a planetary throwing out at human hands.

The loss of these great green reservoirs of life was happening at so fast a rate in Costa Rica that the nation, in very recent years, has undertaken to do something about it. As a result of enlightened planning and leadership, a system of national parks was founded in 1970. It had to be conceived and designed in large terms. The parks were not to be limited to sanctuaries, or isolated islands of life, but planned so as to contain areas of considerable size where native species could be free to develop and would not die out through lack of contact and genetic exchange with the surrounding world.

Costa Rica is a land of great diversity. You can stand and shiver in the high zone of the *paramo*, at an elevation of some eight thousand feet, a tundra-like region of stunted growth, inhospitable enough to be associated with *la muerte*, or suffer from the heat in the dry grasslands or swampy areas at sea level. The parks, designed to reflect that diversity, include offshore islands with nesting sea birds; coral reefs; beaches, such as the one at Tortuguero where the sea turtles come in to lay their eggs; mangrove swamps; marshes rich in water birds; grassy savannahs; intermittently active volcanos; and forests ranging from moist to wet, whose various zones, at different elevations, contain species that are particular to them.

As it has in so many parts of the world, development, out for short-term gains, regardless of long-term results, often threatens the very existence of the wild and all its adventurous species. The exchange in space that is so vital to them is increasingly endangered by the Great Exchange Robbery of our times. It is no easy matter for a small nation, suffering from population pressures and incomes close to the poverty line, to resist. Since development for "emerging" countries is an ideal, it is difficult to defend conservation unless it seems to lead toward economic and social advantage. Conservation as an alternative to development may only look like another form of exploitation if it is seen as taking food from people's mouths, or money from their pockets. Yet the new program of setting parks and reserves aside, if it has not always had universal support, now appears to have gained growing understanding and acceptance. The people of Costa Rica, after all, have seen much of their land slide down hill, and its plants and animals disappear, and have felt the world grow emptier as a result.

The parks were designed to protect the nation's water supplies, pure water being one of the real wealths of a country not rich in the kind of exploitable resources that talk the loudest in our world. They were also meant to encourage scientific research and public education, to establish reserves of native trees for reforestation, and to help in rural development. This national effort, which took a great deal of determination and common sense to bring about, has resulted in parks and biological reserves that comprise eight percent of the land area of the country. New acquisitions may bring this to ten percent, an extraordinary figure for any nation, especially one plagued by economic uncertainty. Proportionately, it far exceed the investment of the United States in national parks.

This commitment to conservation is directed toward a people's attention to their own land and a recognition of it in themselves. The soil is the man, woman, and child, and the trees their fellow guardians of earth's identity. Five centuries ago, Leonardo da Vinci put the equation directly enough:

> Nothing originates in a spot where there is no sentient, vegetable, and rational life; so that we might say the earth has a spirit of growth; that its flesh is the soil, its bones the arrangement and connection of the rock of which the mountains are composed, its cartilage the tufa, and its blood the springs of water. The pool of blood which lies around the heart is the ocean, and its breathing, and the increase and decrease of the blood in the pulses, is represented in the earth by the flow and ebb of the sea, and the heat of the spirit of this world is the fire which pervades the earth.

CHAPTER 18

Sacred Places

U nder the heading *sacredness* there are several references in the Oxford English Dictionary to real estate and property, which might reflect the religious yearnings of our economic system but probably has more to do with the rights its holds most dear. By contrast, the sense in which living things, plants, animals, mountains, springs, and forest trees, were once held to be sacred and associated with deities and ceremonial observances has been cast aside in favor of another, many sided deity that has a tendency to crush its worshipers. No compromise has yet been reached between the scale of modern expansion and the worlds of nature. They must survive as best they can. It amounts to an imbalance that may succeed in destroying several million species and turning major parts of the world into deserts.

Human societies have never been able to conserve and stabilize their environments for long periods of time. It is as if we were not only ignorant of how not to let things go too far but had an often uncontrollable desire to commit sacrilege. Is it not typical of us that since the sea is a symbol of the infinite we should want to ravage it? This goes with a subservience toward authority that tends toward disastrous rigidity. It is a wonder that the earth has not shrugged us off by this time as a major irritant.

It has to be said that the Western world, in attaining its new power to migrate beyond its means, has seldom been at peace with itself. We are split between homocentric power and our origins in the world of life. While the universe endowed us with thought and consciousness, terror, greed, and a haunting sense of the shortness of existence go with it. How can the life of the planet and such a dominant but unruly creature coexist? When there were fewer of us, with less overwhelming power in our hands, we could attain an equilibrium between the earth and the human spirit, changing, shifting with the weather, but understood. We belonged within the earth's orbit, poised between light and darkness, and our senses could equate their risk or well-being with the other living things around them. Without that relationship we do not know the earth, and when its species go into extinction because of us, we have little in ourselves to protect them with. When nothing is sacred, nothing is safe.

(The extinction of species, now proceeding at a rate never before encountered in the history of evolution, is often ignored, in part because "we can get along without them," and in part because the idea is unthinkable. Extinction is anathema to most spiritual beliefs and unknown to children, as well as birds, which, if their sense of being could be translated into words, would probably declare their immortality, in keeping with nature's principles. It is interesting that two of the better known dictionaries of quotations have only one reference each to the term *extinction*.)

Many thousands of visitors to Costa Rica have seen the resplendent quetzal, *Pharomarchus Mocino*, a bird that is doing well in protected areas of that country but has a precarious status in other parts of Central America because of plume hunters and the destruction of the forests. While we outsiders see it as exotic, or something to be added to our life list of birds, the quetzal was sacred to the Maya and the Aztec. It is said that the Maya never killed the bird for its plumes but plucked them while it was alive, which had less to do with what we call conservation than with a belief that what was sacred could not be violated. Its beautiful golden-green feathers were reserved for rulers. In *The Ancient Maya*, Sylvanus Morley described the headdresses into which quetzal feathers were woven:

> The framework of these was probably of wicker, or wood, carved to represent the jaguar, serpent, or bird, or even the heads of some of the Maya gods. These frames were covered with jaguar skins, feather mosaic, and carved jades, and were surmounted by lofty panaches of plumes, a riot of barbaric color, falling down over the shoulders.*

On the assumption that their Spanish conqueror Cortes was a god, the Mexicans sent him splendid presents, including headbands of quetzal feathers and cotton cloth embroidered with them; and a cruel, deceitful god he turned out to be. Montezuma, forced to fill a room full of gold for his benefit, is said to have sadly protested that he would give it all for the feathers of a quetzal. Perhaps that was the point at which money began to degrade the New World environment, or at least take its visibly godlike qualities away from it.

The ancient god Quetzalcoatl was the rain god of the Toltecs, a deity associated with peace and plentiful harvests. Among

*Sylvanus G. Morley, *The Ancient Maya* (Stanford, Calif.: Stanford University Press, 1946).

later people he was the god of life, of wind and the morning
star. His name is derived from *quetzal* (the Aztec *quetzallo* means
precious, or beautiful) plus *coatl*, the Nahuatl word for serpent. On his back, in the culture of the Toltecs, he wore the
brilliant plumes of the bird, and he carried a staff in the shape
of a serpent. (Alexander Skutch, the naturalist of Costa Rica,
has suggested that the union of these two eternal enemies, bird
and snake, might be equivalent to making a single deity of God
and Satan. The Toltecs may have been symbolizing an end to
strife and the beginning of peaceful existence. In any case, the
uniting of disperate elements into an ecosystem, as we might
put it, seems like far less spacious a concept.) The word *coatl*
also meant twin brother, which added still another dimension,
combining, as the planet Venus did, the twin nature of a star
of the morning and of the evening. So Quetzalcoatl came to
mean the god of the morning, and the evening god was his
twin brother Xolotl. What greater part could a bird play in the
religion of a people!

The first quetzal I saw, on one of several trips to Costa Rica,
was in the cloud forest reserve of Monteverde. It looked exceptionally tall and regal as it perched quietly near its nesting hole
in a tall tree, with a long, curving train of plumes hanging
down below the tip of its tail feathers. The green feathers on
its back were iridescent, like spoils of light, and its crimson
breast had all the openly defiant qualities of blood in the reflected
rays of the sun. From the wing coverts, the most elegant green
feathers curved over its breast on both sides. On the head was
a crest of brushlike feathers that reminded me of ancient, martial helmets. The two very long, slender plumes looked like
too much of an adornment for a bird that nests in tree cavities
to handle. But, in fact, the male, when he takes his turn sitting
on the eggs, faces outward, while the plumes are folded over
his back, their tops often seen waving gently in the forest airs.
"Solomon in all his glory was not arrayed like one of these."

The quetzel is a shining facet of the great civilization of nature, where the spirit of human life was once inextricable from birds and flowers and tall trees rising from buttressed trunks with branches smothered in bromeliads and epiphytes, a context of growth and sacrifice reaching through intricate shadows toward the sun. In an open clearing at the edge of the forest where the quetzal and his less extravagantly adorned mate were nesting, a wattled bell bird called with a loud, single "bong," which sounded less like a bell than a metal pipe being hit by a hammer. Inside the forest, nightingale thrushes hauntingly sang, like fine instruments being tuned up to some ineffable scale; and the last I saw of the quetzal was a shimmering waterfall of color plunging down off a branch to disappear in the darkness made by endless leaves.

To think of the dark and tenacious rain forests in terms of the diversity we say is necessary to natural systems is useful to the conservationist, but it is not enough. We who spend our lives guided only by terms and categories, endless facts and numbers, have not yet recognized the depths that would, if they could, help us out of our simplicity, the lack of diversity in ourselves. The great tropical message is inclusion. The forests, with their endlessly varied functions and differences in form, are statements as to the total involvement of life. They are the original grounds of life's inventions, a great drawing in of all kinds of possibilities, over endless time. Without them, we lose not only their incomparable species but the foundation of shared existence.

There is one great, constant element that is seldom, if ever, brought in when conservation is discussed, and that is the night. In spite of the fact that it is so common and inescapable, we have confused it with finality. We have hidden from it, tried to exclude it or cut down on its immense scale, profaned its majesty, called it bad names. That its immeasurable beauty should be included when we think of the value of lands, forests, and seas may not occur to us, but it is in them that the whole night is weighed, contained, and experienced. To lie down with the earth and deny the night is to lose what stature we gain through sharing it.

In the southern part of Costa Rica, in the Talamanca mountain range, is a magnificent new national park of four hundred eighty thousand acres called La Amistad. The name means "friendship" and comes from the original plans for a joint effort between neighboring Panama and Costa Rica. On its side of the border, Panama has not instituted any corresponding park lands, possibly because it imagines it has more important ways to use its money. In any case, I traveled to La Amistad with a group from the International Division of the Nature Conservancy, out of San Jose. It took most of the day to reach the park, and the roads leading into it were made slippery by showers of rain, red tropical earth now churned into ruts as deep as chasms.

That evening, fireflies ferried their very large, pulsing lights all around the clearing where we spent the night. At the base of the surrounding hills and the mountain slopes that rose beyond them, they were winking like campfires. After dark, a large brown and white moth visited the bunkhouse where we slept and lighted on the table. The design on its wings looked as if it had been taken from a native African basket, suggesting that all nature carried the crosscurrents of culture in herself. As I walked out into the night, I heard the waters of a mountain stream rustling by, and I felt the stars and the trees moving with the earth in its orbit, a beautiful interdependence. How wonderful it is to be part of the present mystery, the depth of a darkness never to be explained, out of the universe eternally timed. You pray never to be so far removed from the night as not to be received by it, this magic greater than all our fears.

In the early morning, when the stars were retreating, flocks of little green parakeets began to fly from one leafy tree to another, sounding like squeaky wire wheels. We set off on horseback to climb up to seven thousand feet or more in the mountains, riding cautiously up and down steep ravines along the way and forging rocky streams of purest water. On the lower slopes we had passed an Indian burial ground with one tall stone still standing in it, sacred death in the depths of the forest, where a tinamou, or *gallina de monte* sounded its tremulous call. The now displaced Indians and their lost culture played back the music of the birds with their flutes. To them, the jaguar, hawk eagle, frog, turtle, spider, or lizard were elements of the world of the spirit, and when they died they took these soul symbols with them in the form of golden ornaments. A totality from outside has long since made a clean sweep of those customs and beliefs; but the magic remains, in a beetle with a brown, varnished shell crossed by gold zig-zag markings, or the call of a thrush heard rising upward through the green sides of the mountain.

We had a cheerful, good-hearted guide named Salomon Romero, half Indian, who had spent much of his life in the *bosque* (wood, or forest). He had a family feeling for it. In fact, from the way he would transplant a displaced seedling on our route, such as a high altitude cedar, or a bromeliad, he seemed to treat it as his immortal garden. As we rode slowly up those slopes, past tier on tier of columnar, climbing trees, I thought that the tone of the forest was not of warring, predatory elements so much as an intensity of peace.

In the park that includes the volcano of Rincon de la Vieja (roughly translated "the old woman's corner," presumably by the chimney) evening's blue haze was settling in over the plains below it, while flocks of birds arrowed away overhead and dragonflies darted stiffly over open ground. Smoky clouds gathered over the summit and pit of the volcano, vultures glided toward it and then moved away. From the forested slopes came the shuddering roar of howler monkeys, and the clear, in-shadowed tones of forest birds. We also heard the calling of a quail, and since the wind was from the north, it was said to be a sign of good weather. After dark we found a group of luminous toadstools shining like the lights of some tiny, faraway village.

In North America, we are influenced by the extremes and conflicts in a weather system that ranges over many thousands of miles. The space we occupy is governed by storms from the Gulf and the winds off the glaciers, which may be one reason we can't keep still. Tropical rainforests seem calm and contained by comparison, growing in unchanging constancy. Each life seems not only to take on the characteristics of its surroundings but to be able to change with them, almost mercurially, like the brownish lizard we saw laying eggs on a patch of bare ground. Coming back to the same place a few hours later, we found it on the branch of a small overhanging tree; its skin was now dappled to fit the leaves spotted with sunlight. There were also moths to be found by daylight that mimicked wasps, so as to discourage predators. There were bats that pollinated plants by night, and I heard of a nectar-eating mouse that pollinated a flowering shrub. No relationship seemed impossible, and, for that matter, no size, after I saw a caterpillar of a nocturnal moth that must have been at least six inches long, looking like a stubby cigar. There were also walking sticks of an astonishing size, as well as giant grasshoppers, ants, and cockroaches in various areas. Stout cicadas in the forest canopy sounded like greatly speeded castanets, and they would die down and start up again, with the long rhythms of breath. Endless transmutations seemed commonplace alongside rivers that ran like paths among the stars, smoke from volcanos that joined the exhalation of clouds, hummingbirds that splintered the light from the sun.

Fifty kinds of hummingbirds have been found in Costa Rica, and some of the names given them reflect the glittering beauty of those vibrant, tiny beings, such as the scintillant, the violet sabrewing, the firey-throated, the long-billed starthroat, and the purple-throated mountain gem, which I saw hovering next to the pink and white flowers of a shrub at a high elevation. Some tropical birds have such splendor in their plumage as to seem beyond need in any rational sense, although a quetzal, or a scarlet macaw, only displays the latent fire that their world sustains. Out in an open hillside one day I saw a long-tailed silky flycatcher as it perched on the very tip-top of a tree in the wind. Its feathers were of an elegant shade of gray, with light yellow markings. Trim and perfect it was, as soft as flowers. Then there were the butterflies, more than on all the continent of Africa in this tiny nation, of a dazzling variety of pattern and color I never knew existed, fast-flitting, "vadegating" flyers I chased with my camera and hardly ever caught. I also remember a silvery beetle in the hills that was like a ship at sea under the light of the moon. These are the forms of a supreme creativity, and any country that still has the means and the will to identify itself through them may still survive despoiliation by forces beyond its control.

My fellow hikers and I were sitting on the crumbly lower slopes of Rincon de la Vieja. Jets of steam came out of small depressions and holes in the area, and their gases smelled strongly of sulphur. Yellow crystals rimmed holes where a liquid volcanic mud was boiling and bubbling. The ground was colored iron red and gray, pink, ochre, and yellow, and there was something about this thin surface of the planet's fiery womb that was curiously mesmerizing. We lingered there, sitting between the heat of the rock, rather than by the colder stream that ran nearby, as if we sensed where our deeper affinities lay.

Flying in by light plane to the national park at Corcovado on the Pacific coast, we passed over wrinkled and folded mountain slopes, brown as well as green in the dry season, here and there smoking with fires, the round clumps of spinach-green trees looking like part of an architectural model. Through intermittent cloud layers and scanty showers we finally saw the white brushstrokes of waves along the shore. The little plane landed in a grassy clearing, a narrow strip of bumpy ground, with high trees roped by vines on either side, while overhead, elegantly cut and plumaged swallow-tailed kites made tight turns and easy sweeps through the heated air.

When I first walked the trails at Corcovado, the rain forest seemed like a great, self-perpetuating wheel of light and growth. In spite of intermittent bird calls, the barking of monkeys, the sudden crashing of leaves where something moved, or gusts of wind through the clearings, it was full of an underlying calm. Its plants and animals seemed to attend on each other with an abiding patience . . . I think of those lizards with throbbing throats, in their perpetual suspension.

I heard of the jaguar, whose muscular body flowed with the grace of living water as it chased and killed an agouti or peccary in the forest. I watched orange-brown spider monkeys as they traveled through the leafy, open spaces between them with an easy, reckless freedom. When gaping human intruders stood below them they would shake the branches so as to knock down debris and try to drive them away. The bird calls from one part of the forest to another seemed to signal the stations of diversity.

Stealth and attack were waiting in that multiple containment. I saw several giant damselflies, whose gauzy wings, several inches across, looked like water surfaces dancing in the light; they were almost invisible except for a light blue or sunlight-yellow spot at the tip of their wings. This insect breeds in species of bromeliad that collect small pools of water, and it robs spiders of the food they collect in their webs, zipping in quickly and lightly to snatch it and carry it away. I have never seen such brilliant dexterity.

I stood cautiously aside as columns of army ants set forth from their bivouac in a hollow tree on one of their periodic raiding expeditions, like "armies of unalterable law," devouring other insects, such as crickets and katydids, along their line of march. Many thousands of them streamed ahead, and the prey they drove before them, often escaping to the side, attracted birds to dart around and snap them up.

Then there were the ants that occupy acacia trees, in a sort of protective intimacy. The pioneer females cut out holes in the green thorns of the tree, hollow them out, and then lay their eggs, which hatch out into workers. Colonies feed on nutrient bodies in the leaves. They will attack any predator that tries to climb their tree, and they will even kill off competitive seedlings at its base that might take away its light. It has been found that in Nicaragua acacia ants will prune off the tips of the tree so that the crown grows into a kind of thorny hedge, a protection against predatory birds.

I am a watcher at a distance, filled with awe, trying to add to my understanding of these marvels, through information and knowledgeable guides. It is a matter of digging below the surface of our curiosity, and we can never dig deep enough. What does purpose mean that seems so demonstrable in these ants with their miniscule brains? Am I only to put down their behavior to nerve ends and chemical pheromones and let it go at that? How can we apply comparative values of intelligence to them if we are prejudiced in our own favor to start with? As for me, I think I live half in and out of dreams. We do not own intelligence; it is an attribute of the planet, together with all the fine degrees of perception and awareness in living things, so close at hand in the rain forest, that supreme architecture of life in space. The psychic receiver in us, the biological inheritor, is ready for each scent, sound, and motion within the trees. We have been here before. The life that is nourished and consumed at the same time, that grows its extravagant forms in all their means of escape and magnetic attraction, is still our basic sustenance. It is our inescapable origins that deserve a sacred name.

Out beyond the forest edge at Corcovado was the warm Pacific, with a dazzling sun by day firing a long, gray beach, where sea turtles moved slowly up at night to lay their eggs, often to have them dug out by local people, or predators such as dogs, pigs, or coati-mundis. What little ones do hatch out are preyed upon by ghost crabs, gulls, and frogate birds as they run the gauntlet to open water, where they are threatened by sharks and predatory fish.

At night the dazzling jewels of the Southern Cross climbed their vaulted ladders, and the planets shone like glow worms, or bioluminescent organisms in the sea; while in the forests the animals listened, and the plants stirred in the great silence sanctioned by plenitude.

This procreant ceremony with its guide in the stars will not fail in its supreme timing because of the human race and its mindless destruction. We are its dependents. Night and the sun bring us to justice. One very hot day in the tropical dry forest at Santa Rosa, where the trail ran through a sparse growth of shrubs and trees and was hard to find again if you left it for any distance, an elderly woman in another group I was with wandered off and was lost. It took three hours to find her, now almost overcome by heat prostration. It just so happened that the old "norte Americano" had pinned a sign on her back since she was slow, which read: "Don't follow me." In the ambiguity of fate, that might have been interpreted as good advice. She survived, in part because of common sense and her failure to panic, but this was a hint of the end of the line, as far as you go, although nothing really knows how far it goes. What is death, after all? I only know it through having lived. If I feared it, it was only because I was afraid in life. Under that tropical sun death seemed almost casual. What a common thing it is to lose your way and fall prey to that burning sun which provides all limits, joining a desert we have never seen. This living, this dying, shared with every other form of existence, including fellow parasites and predators, some, like the insects, only living for a few hours, is not of our making or determination. Duration is nothing, association everything.

CHAPTER 19

The Art of the Ultimate

Scientific insight into the universe goes further than our ability to take it in. Its view goes out beyond the world we know, in terms of unassimilable periods of time. In theory, too, the human race can go anywhere, improve on nature, and learn to take action on any of the problems it has brought about. At the same time, space has been confined on a global scale. That freedom of space which called the ships out over the round world was followed out to the extent that whole nations now find themselves in the same murderous room.

The view from near to far, in the days of rooted localities, has now reversed itself. We have an overview of the planet. Local identity has to be rediscovered, replanted, and reinvented. When natural resources begin to thin out and disappear, we do not necessarily feel the loss in ourselves, as we would if they were equated with our everyday lives. If the links in the fabric of the living world begin to fray and part, we hardly know enough to restore them.

Still, there are those immutable standards that lie in the transformations of water into ice, ice into water, and there are mysterious recombinations and revolutions going on below the surface of life that we can only sense in our unaccountable selves. Behind the world so recklessly and uncertainly claimed by politics and economics lie the magic and inexorable laws of the wilderness, known to every life. The flower is wiser than the machine.

The earth's regional distinctions can now be viewed in the living room, although we may not know the name of what it is that grows or flies outside the window. The image of the Antarctic is superimposed on the plains of Africa. Where now is Ultima Thule, once thought to be a frozen zone beyond knowledge or reckoning? *Thule* meant "darkness" among the Phoenicians, a "shadow" in Tyrian speech. It was the terrible abode of eternal winter, where life could not exist. Now you can fly there in a matter of a day or two, or even a few hours, depending on how far from it you happen to live. It is even possible, given the money, to be landed on the North Pole and then retrieved before it gets too uncomfortable.

On a writing assignment to Greenland that only took a few days over two weeks and was accomplished by boat, a small, refitted ferry that once ran between Oslo, Norway, and Hamburg, Germany, I only skimmed the surface. At the same time, that massive, silent land was an opening out for me, a magnification of what I had left at home. Floating ice, although spectacular, was no less so than the winters I had experienced, and the gradual diminution and disappearance of trees I witnessed en route to Greenland began on home shores. Wherever we go, we have polar examples waiting in us for recognition.

Greenland was a vast space, filled by single images: two men in an old dory with an outboard, heading out on a dark day carrying an antique rifle with them; the round body of a ringed seal lying in the water by a dory; a minke whale being cut up on a rocky shore; a black and white razorbill in the hold of a fishing boat; plus the longest day, the longest night, a spare world thrown open to titanic beauty, a land with no disguises. It is still in the ice age, so that most of it is covered by a sheet of ice, or compacted snow, and habitation is narrowed down to ribbons of land along the east and west coasts.

There are now forty thousand Greenlanders, of which some nine thousand are Danes. Before the country became a Danish protectorate, beginning in the eighteenth century, it was occupied by the Inuits, a hunting, fishing people whose population may have been somewhere between six and eight thousand. In the capital city of Godthaab, the trucks roar and the winches of the cargo ships on the docksides clank as they unload concrete slabs for new housing units. Row houses, sad and severe, face the outer waters instead of sod huts. What were originally isolated trading stations have expanded to include department and supply stores, and the necessary hotels and airstrips, but highways do not exist.

In spite of a social security net provided by the Danish government and a continued effort to strengthen and encourage the local economy, neither modern organization, technology nor money can quite sustain the balance between a larger population and what is required for its subsistence. The Danes were conscientious rulers, anxious not to overdo their control and to steer Greenland toward self-determination. Many of them have intermarried with the Inuit Eskimos and now call themselves Greenlanders. Home rule has been instituted, but there is dispossession in the air. The unemployment rate is twenty-five percent, those whose displaced culture never included alcohol now take to it as a refuge.

There were once enough seals to feed a few thousand people, but not forty thousand. New shrimp plants and expanded fishing industry try to fill the gap, but periodic scarcities are as much a part of life as they were in the days of the skilled hunters, when cycles of starvation and plenty were unavoidable. In recent years, an already short summer season, with its ice-free waters has shortened by several weeks. The catch of codfish has declined because of colder temperatures, as well as overfishing in Greenland's rich grounds by foreign nations, which have taken hundreds of thousands of tons each year.

I myself carried my demanding twentieth century baggage with me and saw iron limits facing it all around. This, I felt, was a place that would never be so possessed by us that we were unable to distinguish it from what we did to it. You look directly out on the great overarching bounds of nature without intermediaries. Civilized complexities fall away before a land that is so magnificently complex in its simplicity.

In the smaller settlements, with their little green, red, and blue houses perched on dark rock, surrounding harbors filled with fishing boats, there were sled dogs tied up everywhere. that would set up choruses of yips, snarls, and howls. Otherwise a silence dominated, as it did over most of Greenland; and we transplanted Americans walked the streets, wondering where the action was.

The statistics are staggering. The ice cap, or ice sheet, covering a major part of the island—the largest in the world—is up to thirty-five hundred meters thick. If this ice were to melt, the sea level of the world would rise by some six and a half meters. In a number of areas, glaciers move down from the ice cap to the sea, and icebergs calve off them with a thunderous roar. Where the icebergs mass together offshore at Jakobshavn, they are called ice mountains, millions of tons moving slowly but measurably along, at about thirty meters a day. Their towering, white crags are shaped like cities of the imagination. The Great Wall of China is a minor thing by comparison. You look at constructions that existed before our world began, but lived with, watched, for thousands of years by settlements of superb Inuit hunters, who traveled out between them in their kayaks in search of seals.

On the west coast there are mountains estimated to be between three and four billion years old, close to the estimated age of the earth itself. There is clear evidence of two or three glaciations covering these mountains, and as many as twenty more may have occurred over the past ten million years. They wore down the mountains by as much as three hundred feet, and they look about as worn down, scoured, and scored as any you are likely to see. How is it possible to conceive of a billion, or even a million, years? You can hold it in your hands. I did not get much of a thrill out of picking up a piece of the rock. I confess it felt like any other, and being no geologist, I was unable to analyze its characteristics. But if, through handling it, I became a part of global upheaving, wearing down and rebuilding, of violence, pressure, and change, of ages beyond comprehension and a present made of no finality, then I could heave it at the ice with real abandon.

Rock to the Inuits before *kablunaat*, the white man, came, and they were "still strong," was part of the central games of life. Ulli Steltzer's book, *Inuit*, quotes one of the people, Jimmy Muckpah, as saying: "Inuits make *inukshuks* (rock figures or cairns) all over the land. They did not have CBs and radios. If there was good fishing, they put up rocks to tell the next guy where to fish. If there were caribou, they made big *inukshuks* to hide behind. When the caribou came closer, they shot them with bow and arrow."*

Playing games with rocks, from childhood to the grave, in that monumental landscape, how could they possibly separate themselves from what we call the "inanimate"? In putting rocks

* Ulli Steltzer, *Inuit: The North in Transition* (Vancouver/Toronto: Douglas & McIntyre), 1982.

apart from us as if they only meant what a mineralogist or a rock crusher can tell us about them, are we playing a misguided game? (I was told that boys who once practiced the skill of their forefathers in the use of a harpoon by casting stones were now, "unfortunately," throwing them through windows.)

At the outset of our trip, in the tundra surrounding the airstrip and itinerants' hotel at Søndra Strømfjord on the southwest coast, I saw my first caribou. I walked off for several miles over hills covered with heathlike plants, cloudberry, arctic willow, and dwarf birch, with the blue flowers of hare bells nodding in the wind, to see the edge of the Greenland ice cap, a long rough wall in the distance. I found some caribou tracks; and then it suddenly materialized, a clean-limbed animal standing and watching me, the Arctic in person. We stood watching each other for a few seconds, then he thrust back his head, with its load of antlers like candelabras, and took off at a fast trot, a beautifully swinging, graceful gait. I saw several more caribou that morning, browsing in the distance. I also came on several piles of whitened bones. Since hunting was not allowed within twenty miles of the base, these were apparently caribou that had died during exceptionally severe winters, when the ground was so deeply frozen that they were unable to get enough to eat.

A dark little arctic fox loped gently through the low undergrowth, taking an erratic course. Snow buntings and lapland longspurs flitted over, while black ravens conversed in gutteral tones on the ground, sparring lightly with each other; or pairs would rise up and play together in the air, dropping down, gliding to the side. It tempts you to try and interpret raven speech. One summer, while my wife and I were picking blueberries in Maine, a single raven sat in a nearby tree, and I could have sworn that it was making ironic, sardonic comments about us. The Greenland ravens were very talkative, making a variety of palate-fluttering sounds, at times high-pitched and nearly warbling, along with their "kaarrh," "kruk," "rahwk," clickings, and so on, while they were never still, their wedge-shaped tails constantly twitching. It is a playful, crafty, confident bird and, on that ancient, gray-green hillside, seemed in complete command.

Iceland and glaucous gulls of an almost pure white, or with wings tinted a very light, almost diaphanous gray, cruised around the harbors, or flew over rough water farther out, their wings having that supple, bending assurance of all the race of gulls. As our boat followed the mountainous coastline, it was accompanied by thick-necked, stiff-winged fulmers, gliding over the surface hour after hour; the rougher the weather the more they seemed to like it. As they planed over calmer waters, their wings seemed to caress the silvery sheen of the surface, barely touching it as they sped by.

"A great piece of engineering like that," said a fellow passenger, "is bound to survive." That engineering, an accommodation of form to the immeasurable, kept gliding on, beyond our grasp. The mood of the weather continually changed. It snowed on the twenty-fifth of August. Strong winds and rolling seas would shift to an almost glassy calm later in the day; snow squalls suddenly drove in, and then the clouds cleared away and the sun shone. The unpredictability of the weather played havoc with schedules, grounding or delaying planes and modern travel in general. Great walls of cold were moving closer. I felt incapable of living here, of measuring up to this land; at the same time, it was all I ever hoped for. In gathering darkness, at ten or eleven at night, magnificent icebergs rode out in the fjords near their junction with the sea, gathering the subtle changes in the light, catching green and gold against their lined and glistening sides. You are not prepared for the sheer glory of the Arctic.

In the Disko Bay area, a glacier moves out into seawater

from the hills, with the ice cap that spawns it showing like a band of white clouds in the distance. We approached it at the end of a long, deep fjord. There was a cannonading sound, and out of the frontal mass of the glacier fell an enormous chunk of ice into the water, making a high white shower and sending off a shock wave over the water. The upwelling and its violent pressures stunned or killed great numbers of fish, which attracted hundreds of dainty kittiwakes. There were so many of them feeding together that they looked from a distance like a mass of great insects gathering over a carcass. The sea was glassy smooth and the water a stony-blue with icy scales over its surface. As the boat plowed ahead, crackling pieces of ice broke off and danced away like skipped stones. On one side of the fjord, the low mountains were almost without vegetation, lined faces of gray rock, while on the other the ground was covered with reddish-brown plants of the tundra, and all the surrounding horizon kept shifting color, from pink to purple to blue, as the sun shone through a hazy atmosphere. I felt a great suspending in the place, like the tidal seas, and took away a returning awe at one of the great sights of the world.

We spent one afternoon slowly moving between the sculptured icebergs, gleaming giants, their sides scoured and shaved, and with veins of a brilliant cerulean blue. Where the waves chucked in along their eroding flanks, they reflected a dazzling blue-green. They were carved in the shape of flames, cut through by arches hollowed out with caves. Their sides flared steeply toward the sky, or they were rounded, the result of shifting and rolling over. Great chunks fell from them with a sound of thunder, or there were sharp cracking sounds as we sailed by. When the sunlight hit them, their sides had a soapy sheen.

"Henceforth", said Emerson, in *Nature*, "I shall be hard to please. I cannot go back to toys. I am grown expensive and sophisticated."

Peter Freuchen, friend and associate of the explorer Knud

Rasmussen, told a story of climbing to the top of an iceberg and finding a cave where there was a pair of polar bears. He lured them out with a rag fastened to the end of a harpoon, and they both emerged, at which point "Knud fired."

> There is an old suspicion among the Eskimos that much evil will result from skinning and cutting up an animal on an iceberg, but there was nothing else to do as the bears retired into their cave to die, and it was impossible to get them out whole. The blood flowed like a brook and disappeared in a crack as he cleaned the animals. We hooked our lines around the carcasses and began to pull them out of the cave.
>
> The moment we heaved on the meat there was a detonation like a cannon shot . . . the thunder rolled on and on. I felt as if I were treading air, and saw Knud tossed high above my head. Then I knew nothing until I saw Navarana (his wife) smiling down at me.*

The iceberg had exploded, and they were hurled from it and out onto the ice below. It had toppled over and was now bottom up, but no one was hurt. According to Freuchen, the explanation was quite simple: "Icebergs are formed under terrific pressure within the icecap. When they slide out to sea and this pressure is removed, the slightest thing may alter their balance and throw them into a violent readjustment. The warm blood of the bears flowing into the crack had done just that. As is so often the case, unreasoning superstition derives from an intrinsic truth."

Who could imagine anything so extravagant as an iceberg, perhaps two hundred feet above the surface and nine hundred below, being turned over by a stream of blood! Yet it strikes a familiar chord. The human soul senses a balanced immensity in the world, combined with an extreme delicacy of response.

* Peter Freuchen, *Book of the Eskimos* (Cleveland: World), 1961.

Who knows what we might do through some random, ignorant act so as to turn the mountain over and knock us unconscious on the ice below?

Anything I could think of to say about those high ice beauties was bound to trail them, a long way behind. The words I used ought to be measured against what I was unable to do, such as carve an iceberg, create a gull, everything that gives life its completeness up to now, its stature. We are always found wanting.

Knud Rasmussen, born in Jakobshavn, was a strong-willed, joyous, and courageous explorer of his native land. He did more than just beat polar records with his dog sleds. He recorded the songs and folklore of isolated hunting people before they disappeared. He said of them that they possessed a "humility in the face of the pressures of life which springs from an innocence for which we must surely envy them." A material people, constantly facing sickness, hardship, and death, they could also sing and dance in an ecstacy founded in the elemental truths of their surroundings. Their "spirit songs" were not conscious art as we know it but an expression from the depths, a crying aloud to the empty air.

Rasmussen quoted a Netsilik man by the name of Orpingalik who told him:

> Songs are thoughts which are sung out with the breath when people let themselves be moved by a great force, and ordinary speech no longer suffices.
>
> A person is moved like an ice-floe which drifts with the current. His thoughts are driven by a flowing force when he feels joy, when he feels fear, when he feels sorrow. Thoughts can surge in on him, causing him to gasp for breath, and making his heart beat faster. Something like a softening of the weather will keep him thawed. And then it will happen that we, who always think of ourselves as small, will feel even smaller. And we will hesitate before using words. But it will happen that the words that we

will need come of themselves. When the words that we need
come of themselves—we have a new song.

All speech, all the signals of awareness and community in
the living world, originates in the great force, which has no
master.

Tourists on guided trips are treated like perishable foods,
not allowed out, except at stated intervals. We hardly met the
people, although you sensed that their past had not yet lost its
hold, any more than the glaciers had left the mountain passes.
The young Inuit women had inherited the warmest smiles of
any I have ever seen; but the strain to adjust to the pressures
of the modern world were the economy, rather than the person, now tells you how much you are worth, also showed in
many faces.

It seems very strange that I should have come closer to the
people through their dead, although strange things do happen
to those who wander in. The museum at Godthaab, where the
trip ended, had several frozen bodies on display. They had
been found up the west coast at Upernavik, where they had
been buried under rocks. The rock cover had protected them,
winds blowing through it had kept them dry and thoroughly
frozen, and they had been remarkably well preserved. A child,
about six months old, dressed in seal skin and furs, looked like
a doll, with the smooth skin of the face intact, and the hollow
eyes looking as if they still gazed out with a terribly touching
surprise. There was also one young woman lying in state,
clothed in beautifully fashioned furs and boots. It was estimated that they had died, through unknown causes, in about
the year 1475.

I looked at that woman over five hundred years and found it
perfectly natural, as if none of the vast divide between us stood
in the way of basic kinship. There was no "heathen," as the
missionaries used to call them, but the product of a culture

high born in nature. It was a sight sustained by a continuum of myth and mind, of flesh and blood, in love and protective- ness, in the reality of death in life. If only because we were carried together in the timeless flood of existence, I greeted her as my sister in mystery.

CHAPTER 20

Wilderness Flags

The whole nature of the earth is shared by the lives that follow its lead. That is why I would never be separated from a single leaf, whose circulation is traced in the veins of my hand. To what useful end could I use my eyes without acknowledging that they are only one of the earth's inexhaustible ways of seeing? And we have only begun to give credit to the rest of life for sharing mind and consciousness, as if we had been absent from them for such a long time that we had forgotten how.

The lasting wilderness lies out any shore. The trees in your yard were born of an ice age that predates the one we threaten to bring about on our own. The rain that trickles down our necks brings us into inescapable weather. When I walk out over the glistening sands at low tide, following their serpentine ripples to the lip of the sea, I begin to meet with facets of life that rival the night sky in abundance. Even on a cold, obdurate day in mid-winter, I watch the gulls stretch out their necks to give an occasional stentorian cry, or stalk slowly about in shallow water, waiting on another change. Gull nature is inseparable from the following waves, and the conflict between north and south, east and west, that keeps rising on the horizon. On a cold gut level, with fish and gulls, I feel a stirring in the seas. Why talk about the helpless fate of nonhuman lives that are incapable of managing their own destiny, cruelly bound to nature? Looking around me at our "management capabilities," I begin to wonder. Are the birds and fishes not on a sustaining level with magnitude?

Watching and searching is in the air, expectations of the next millenium. I have always felt lifted by the great timing of things, carried out in response to the weights and balances of the atmosphere. That I am only a cipher in the process does not bother me at all. The search and the discovery lies in the approaches to all seas, and in their inclusion and deliverance. Wilderness travelers, everywhere, join all points on earth to every other. Nothing can survive in isolation.

The city I once lived in founded its artificial eyes and its forgetfulness on rock. The glacial moraine I have perched on for many years by the sea rests over deeper deposits, which in turn lie over the bedrock of the continental shelf. The granite slopes I first climbed as a boy are an inseparable part of me and lie under all I have since experienced. Autumn days now open again, many autumns, winters, springs, and summers having flown behind, and I am climbing in the New Hampshire hills I came from. I pass a thrush in the trees, upright, poised in a scared, wild way, wood brown, with a dark forest eye ringed with white, still carrying spring carillons in its spirit, and our two lights meet. That thrush is the apple of my eye.

From an open ledge farther up I see three hawks on their autumnal migration, following the ridges south, slightly rocking, wheeling and soaring in an easy relationship with the high air. The wind and its updrafts carry them down between the valleys between the lesser hills and mountains of the Appalachians. They glide with only a quivering of their beautifully taut wings, taking the misty blue sky on their feathers. So they rock and pass, meeting the wide gaze of distance in their pride. The hawk way is the reliable one, telling me how the continent is measured, mile after electric, flowing mile.

There is a brook pouring down parallel to the trail whose waters carved ages of rock into long, fluted channels, where, as a boy, I used to "shoot the chutes" with free and dizzy excitement, to land in deep cold basins with a great splash. In the eddies and side pools, water striders drift and propel themselves forward with short, thrusting motions, making dimpled reflections on the surface. They meet each other and then spasmodically twitch apart, not unlike birds in flight. The waters of the brook constantly lisp and rustle, rush and whirl, over the accommodating rocks, taking the free way out. Nothing seems to have changed. Even the log shelter where I spent the night on the way up, so many years ago, hearing a porcupine snuffling and chewing on the corner posts, looks much the same. The rocks are stepped and terraced, the way they used to be, and the water goes twirling down the long slides. In some places square rocks are perched on larger boulders like the ruins of an ancient civilization, but this history goes even further back; and it is strange to think that I was a part of it and now return, although everything in my life has changed and I live in an age that never stops upgrading change. I have had all manner of intermediate, often catastrophic worlds falling and rising around me, and no doubt the same process will be characteristic of the next thousand years, while the brook keeps rushing down. Its waters represent our long allowance, and that it accompanies us puts human violence and impossibility in the frame of fantasy they belong to, at least in the context of an abiding space where all abysmal fates are reconciled. Yes, everything has changed, but all the mountain and these running waters have to express is capacity, which in terms of their patience must be far greater than it looks to anyone, any scientist, engineer, or temporary driver with his foot on the pedal.

The brook makes a windlike sound as it rushes past the rocks, shaded by trees that will soon be sugared with snow. What is the function of these hemlocks? Surely it is to receive pale, cold sunlight through glittering needles, layer on layer down to the stillness of the ground. How much we owe to the transmigratory serenity of the northern birches, electric twigs and white branches swaying under blue sky, creaking and singing in the wind as it whispers and roars throughout the mountains! It is going to be much darker in a few weeks, and icy cold, and the blue jay bouncing from the hemlock knows it because it is partly clothed in darkness already. We know jays for their iridescent feathers, for their mischievous natures and their harshly musical cries. But they have other qualities, too, in earth's keeping, as hidden as Mephistopheles. They were born to the secret nature of privation, created by the endlessly evolving nature of planetary energy. Their way of change is as exceptional as the way we think about change. Whenever I see them, or other life forms, in different environments than I am accustomed to, they present another vision, which is masterful enough. Jays are great performers. So it says through the dark evergreens and the glossy birch that stand over the banks and the perpetually running water beating in my ears.

Everything stands in for dignity. I feel the total absence of pretense around me, a power I lack the strength to understand fully. The slow building of icebergs and their melting back, the vast inertia of the rocks, the roaring and dying down of the wind, the enormous lag between what our instant perception feels and ages out of reach, have given expression to the blue jay and the trees. They all seem to say, or stridently cry out:

"On time, on time." And although any man or woman might say, "Not my time," it holds, in the incipience of rainbows, the gradations of failing light, the mysterious sun that ensures a universal rising that nothing escapes.

So I go on climbing, thinking that if my own life is any criterion, the "ascent of man," despite his superior brain, is no straightforward progress from bottom to top. It might in fact go down as well as up, and is at the very least meandering, facing stubborn resistance all the way. I think we were held down for a purpose, so that no matter how far we move out we should return to this monumental gravity, the weight of sea and mountains, the rock that is a real leader to the sky. I walk against the wind and it is in me, a counter force that gives me my direction, like fish swimming upcurrent. That my life is full of dissatisfaction implies promise as much as failure. Incompleteness teases me on, as it was in my beginning, not far from Newfound Lake.

I have no sense of triumph in reaching the top. I have conquered nothing. I only cry out to the distance: "How beautiful you are!" The exuberance of giant shadows lies across it, the waves of a world with all weather on its body, smoldering red, lavender, and dark green, rolling and flaming away, through curtains of mist, with lakes shining down below like spots of glass through a hazy sun. Nothing we do can defeat that impenetrable unity. There is also no final lifting of the veil to tell you where you are, what entitles you to promises. Heights are not climbed to get away, nor to see our lives as the mere spasms they are, but to join in universal application.

Patches of gray-green lichen cover the bare gray rock of the summit, which is studded with protruding knobs of white quartz. Here and there a lichen of luminous yellow-green overlaps the other variety like so many splashes of paint. The rock leans out with long cracks and crevices lining its rounded shoulders, a scoured, endlessly bathed, enduring hide. I hear the tinkling twitter of snow buntings close by and see them rise like a scattering of petals flung up by the wind. They fly out just beyond the summit, then land on the rock again. A light brown face patch appears just over the ridge, and then more of the little birds come into view as they explore rock crevices for the seeds of the tundralike plants that take shelter there. I walk toward them, and they fly ahead of me with gay, airy abandon. I almost follow them out to where the world falls off. Their white underwings shine as they fly, and on the ground they show their wonderfully subtle and delicate blendings of black, white, and a cinnamon that contrasts with a still lighter shade of brown verging on orange.

I know them from the ocean shore, where they dip, rise, and scatter out like flecks of foam across the dunes. Waves are their medium, the waving open tundra, the waves of sand, the tumbling waves of rock from which the grains of sand originally came. They belong to thousands of miles of continental shadows and surfaces. All the way down from the arctic barrens where they breed, they know the contours of the earth. They pick out their chosen place and its fold like a sculptor finding the right marble in the hills, or perfect, water-worn boulders by the shore. Their affinity is to what corresponds to their own genius and distinction as a race. The earth provides, to those who know it.

A number of monarch butterflies are being flung back by the wind, or knocked down, where they flutter feebly on the rocky summit. They are part of an annual migration that is bigger this year, 1985, than any I can remember. Ordinarily, they flit and glide at a not very great height above the ground, and I have been seeing them in town for several days, heading in a southwesterly direction. They get carried aloft by strong winds and updrafts and have been reported at several thousand feet up in the mountains. Their numbers, and their presence, have made me look at them more closely for a change, instead of passing them by as merely familiar. I have begun to see their great orange and black wings as wilderness flags, symbols of a continent. That this butterfly can travel, even at no more than eighty miles a day, all the way to Mexico, seems miraculous. The ready explanations you can think of, such as their powerful wing muscles and exceptionally large wings, do not quite account for it. Again, it is the timing that is hard to grasp. These creatures are halted by cold weather. They die in vast numbers. It seems improbable that enough should survive to return again, although their migration extends from sea to sea, and, like many other forms of life, they have periodic surges in their numbers.

The timing between eggs and larvae in their changes, to the chrysallis, that exquisite jade green vial with its circlet of golden beads, goes deep into the nature of the earth. The known factors that might explain such a phenomenon are surely not enough. The power behind their transformations brings all tides and seasons with it.

The mountain holds a continuity for me, waiting since I was ten years old. I have seen its rocky top covered with a glaze of ice, or hidden by falling snow, times when I could have been lost there, left behind for good, but I have always been reluctant to leave it, in any weather. I stand in the axis of its spirit, like the red-tailed hawk hanging in the heights above my head.

Climbing down through fir and spruce, and then the birches in their arctic white and the bonfire maples, I meet with a nearly immoderate quiet, while scattered leaves fall gently in front of me. I know that the manmade fires out on the world must consume, die down, and flare up again, playing their game of limits with the earth. But it is not human society, but the eternal changes that will determine the outcome.

At the edge of an open field, many monarch butterflies are sipping nectar from a clump of New England asters. They constantly move from one flower to the next, opening and closing their wings. Their slender black legs, their black antennae, like fine filaments, the compound eyes on either side of their head, capable of integrating thousands of separate images, their wings, plus their sense of touch, and other unseen attributes, are enough for their sacrificial mission, but we can only wonder at its terms. Over the billion years it may have taken for a monarch to evolve, life has been throwing its sparkling waves to break upon the shore, invention past invention.

Those wings, with their tiger markings, look like stained glass windows. They are charcoal black, cloud white, and a smoky orange-red, like sundown fires, a subliminal color, taken from the formative nature of creation. They reflect their lord the sun.

There is a seasonal exhaustion in the air. The ground is cool and subdued as the hills turn dusky and purple by late afternoon. I pass cleared fields full of stubble, the lank, dark stalks of corn. Milkweeds, where monarchs deposited their eggs, have opened their pods, and the white silk lies over browning grass like wisps of cotton, or is concentrated in spots like the downy feathers of a chicken caught by a fox. Now summer is broken for good. There is no certainty except in the spread of evening, royal and wide, and in the coming on of a night which all life shares without distinction. Noting less could equal human desire, nothing less could reach beyond it.